LASERS AND ELECTRO-OPTICS RESEARCH AND TECHNOLOGY

HIGH POWER/ENERGY LASERS IN OUR LIFE

LASERS AND ELECTRO-OPTICS RESEARCH AND TECHNOLOGY

Additional books in this series can be found on Nova's website under the Series tab.

Additional eBooks in this series can be found on Nova's website under the eBook tab.

LASERS AND ELECTRO-OPTICS RESEARCH AND TECHNOLOGY

HIGH POWER/ENERGY LASERS IN OUR LIFE

VICTOR V. APOLLONOV

nova publishers
New York

Library of Congress Cataloging-in-Publication Data

Names: Apollonov, Victor V., author.
Title: High power /Energy lasers in our life / Victor V. Apollonov (Head of Department, High Power Lasers, GPI RAS, Moscow, Russia).
Description: Hauppauge, New York, USA : Nova Science Publishers, Inc., [2016] | Series: Lasers and electro-optics research and technology | Includes index. | Description based on print version record and CIP data provided by publisher; resource not viewed.
Identifiers: LCCN 2016019017 (print) | LCCN 2016010043 (ebook) | ISBN 9781634850087 () | ISBN 9781634849760 (hardcover) | ISBN 9781634850087 (Ebook)
Subjects: LCSH: High power lasers.
Classification: LCC TA1695.5 (print) | LCC TA1695.5 .A658 2016 (ebook) | DDC 621.36/6--dc23
LC record available at https://lccn.loc.gov/2016019017

Published by Nova Science Publishers, Inc. † New York

To the blessed memory of my teacher and colleague-
Acad. A.M. Prokhorov

CONTENTS

INTRODUCTION

The book is devoted to the blessed memory of my teacher and colleague Acad. A.M. Prokhorov. I was lucky to work for him for more than 32 years. Even now, I am still amazed by the genius of this great man. On July 11, 2015, when he would have been 99, a great monument was unveiled at the cross of Leninsky and Universitetsky prospects in Moscow to commemorate the glory of this outstanding man.

This book describes the physical features, current approaches to and new applications of high-power, high energy, high-repetition-rate, pulse-periodic (P-P) gas, chemical and solid-state lasers, which we developed together with Acad. A.M. Prokhorov. At the dawn of the laser era, we discussed their ability to be scaled up and all the possible ways of making them more efficient and not so bulky. The book is intended for technically advanced readers who want to familiarize themselves with lasers beyond the general level. Although the book is aimed at students who are thinking of entering this fascinating field, it might also be used by practicing scientists and engineers who are changing their technical direction and want to learn more about this particular type of lasers. After reading the material presented in this book, readers should be able to follow the scientific and technical literature and to understand the physical principles and engineering issues of a new type of high-power, high-repetition rate P-P lasers, as well as to appreciate the subtleties, richness of new applications, and operating possibilities afforded by these new high-power, high-repetition-rate laser systems.

Following the demonstration of the first laser over 55 years ago, an extraordinary number of different types of lasers have been invented using a wide variety of active media and pump techniques to produce an inversion. As a sign of a maturing industry, high-power laser research and engineering have developed into many specialized disciplines depending on the laser medium (chemicals solid-state, semiconductor, neutral or ionized gas, and liquid) and excitation mechanism (optical pumping, electric current, gas discharge, chemical reaction, and electron beam). The development of high-power laser systems represents a multidisciplinary effort and is the result of the cooperation of professionals from many branches of science and engineering. Today, high-power laser systems are very sophisticated devices, and the field of their application has developed so that it is even difficult for a professional to enter it without prior familiarization with the new concepts and technology of this class of lasers.

Fabrication of high-power/energy lasers with a large integral beam cross section by using amplifying channels is a complex scientific and engineering problem. To date, the research teams in the U.S. have designed and actively tested multichannel solid-state lasers with an output level of hundred kilowatts. These lasers are extremely complex devices that require ongoing technical support. On the other hand, diode-pumped solid-state lasers built on traditional optical systems and having an output power below 10 kW, are already quite commonplace devices. The existing solutions make it possible to manufacture lasers, which require virtually no maintenance over their entire service life.

Obviously, one of the most effective solutions in fabrication of high-power/energy solid-state lasers is the multi-channel principle of their design. The laser in this case is assembled in the form of an array of identical channels, each channel ensuring laser generation. The small cross section of the channels allows for efficient and easy-to-implement pumping of the energy into the active medium and heat release from the medium. According to this scheme, a 105-kW laser (Northrop Grumman Corporation) was fabricated. Further improvement of the output parameters of multichannel laser systems can result from a significant reduction in the optical thickness of the amplifier, and therefore from the employment of mono-module disk geometry of scaled-up laser systems. The Airborne laser (ABL) project is another step in the creation of long-range laser weapon systems. But the laser's lethal range was only 85 miles (135 km), which is well below the limit of 600 miles that was initially planned in the projects on missile defense at the very beginning. Military applications of high power lasers are quite obvious but not easy to manage. Laser systems for strategic applications should be very efficient, compact and reliable. As was announced at the Las Vegas 1985 Conference, the range of about 1000 km and more is the strategic one.

In the future, high-power/energy, high-frequency P-P lasers will find applications in the field of ecology. Besides, they will be used to transmit power over long distances, to obtain high-power optical pulsating discharge plasma radiation for arbitrary geometries of the breakdown in a three-dimensional space, to produce very long conductive channels for electrical power transmission and long range shocker, etc.

A new class of rocket engines – a laser jet engine (LJE) – objectively belongs to the most promising rocket engines of the foreseeable future. The LJE is substantially more cost-effective than conventional chemical fuel engines. At the initial stage of the flight, the LJE employs atmospheric air as a working medium, and in outer space use is made of a small space-borne gas supply or an easily sublimated substance. In this case, the specific costs of the cargo launching to outer space may be reduced down to 160–200 USD/kg.

The LJE-based solution of a series of very interesting and important problems is predicted, in particular:

(i) designing a space interceptor that would destroy debris and other dangerous space objects, such as asteroids, meteors, etc.;
(ii) launching of macro-objects with a very high acceleration;
(iii) realizing super-long orbital-scale conducting channels for addressing the global challenges of ecology and solar energy transmission; and
(iv) producing an optical pulsating discharge as a source of X-rays and a wide spectrum of light, shock and sound waves.

A beam of heavy multiply charged ions produced by high power laser seems very promising for a large number of applications. First of all, because of the high charge of the particles, it is easier to accelerate them to high energies. Secondly, it is easier to obtain higher currents, because the greater the mass, the smaller the influence of the repulsion of the ions in the beam due to space charge. Thirdly, the beams of heavy multiply charged ions will transfer the energy to the target with maximum efficiency. In this case, it is an essential issue of how to obtain primary beams of heavy multiply charged ions for subsequent acceleration. A laser-plasma generator of multiply charged ions produces a large number of heavy ions in the P-P regime of short pulses, which is of interest for ion accelerators. The source of this type is also promising for research in the field of heavy-ion fusion. The fundamental idea of a laser-plasma generator of multiply charged ions was put forward 45 years ago with an active participation of the author of this book.

In this book we also outline the possibility of using the above-mentioned high-power/energy lasers not only for removal of dangerous elements of space debris (ESD) from the space vehicle orbit, but also for planned clearing of the satellite orbits from these ESD when the given installations will carry out a role of a "cleaner". Radiation of high-power/energy lasers in this case can remove the ESD from a circular orbit to the elliptic one, which perigee is in the dense atmosphere beds where the ESD should be burnt. As a result of solving the ballistic problem, the dependence of necessary reduction of the ESD speed from the height of their orbit over the Earth is obtained. It is found that for orbits with heights up to ~300 km, the necessary impact can be provided by a DF laser with a telescope with a high enough diameter and a pulse duration of about few tens of nanoseconds.

The technology of high-power/energy, high-repetition-rate P-P lasers is expected to be used for full-scale production of icebreakers equipped with such laser systems. The most effective is the joint application of high-power lasers and icebreakers in the cracking of drifting ice fields covering the Arctic. Another important task for lasers on offshore oil and gas platforms in the Far North will be in high demand and can greatly facilitate the safety of the personnel of such rigs.

So, the future applications of high power/energy lasers in our life are very welcome and look really promising!

Chapter 1

A Talented Person Is Talented in Everything

I was lucky to work with A.M. Prokhorov for more than 32 years and, every time revealing new facets of his many talents, I was always amazed by the genius of this great man. What, above all, comes to my mind about this extraordinary man now that he is no longer with us? His incredibly developed humaneness, his sense of intuition, his striking ability to find quickly the right solutions, and his heightened sense of the new that would be fundamentally significant for a leap into the future. I believe that the feeling of the front edge of science and its development trends are perhaps the most important characteristics of this phenomenal scientist! Anyone who had a chance to work and communicate with A.M. Prokhorov, even for a short time, was blessed with this feeling.

The Institute, at the stage of its formation, was lucky to have a leader like this. Even at the most difficult times, this feeling did not leave those who, despite all the hardships, continued to work actively. The constant state of extreme stress to find the only right solution could turn, by an experienced hand of the Master, into unbridled joy, witticism or a joke. If during a meeting at a seminar one did not burst into laughter at a witty word used to relax the situation, it meant that he did not understand something, that he was not in shape. Loud laughter from the office, from time to time, heard even in remote parts of the corridor, said, "Everything is OK. We continue to move forward. We live."

Prokhorov did not like jokes that were not to the point. His style was the one-sentence joke, touching upon an important issue with a master's hand. Let me remind you of the interview given to NTV correspondent Pavel Lobkov about the causes of failure of Russian candidates for the Nobel Prize. Pavel pondered indecently long over the meaning of the appropriate final sentence voiced by A.M. Prokhorov in defense of V.S. Letokhov. The phrase "he who pays the piper calls the tune" was certainly about the role of USA in decision making of the Nobel committee. Prokhorov's colleagues at the Institute got accustomed to good humor, which was undoubtedly to his credit.

The ability to make a decision, even in an insanely difficult situation when one gives up the fight, is also what he taught us to do. It was important for him to think, first of all, about the common cause rather than about himself, and he wanted us not to be afraid of making a mistake. Any mistake can be corrected, but the time lost to the cause will never come back. A good example here is a series of decisions made during the Perestroika Years. Here is one of them. In the most difficult moment, when the science "was just thrown overboard", it was necessary to quickly interpret the phrase "one can do anything that is not prohibited by law."

"Where to get money for science in order to be useful tomorrow when once again the Motherland will become aware of the greatness of scientific progress?" – That was what we had to think about, standing on the ashes of the former Soviet Union. The solution was simple and effective: freedom was given to departments and laboratories to conduct foreign economic activity via contracts and grants. And it was in those times when neither accounting nor planning departments had specialists for 'shoveling' piles of papers written in all sorts of foreign languages. Several dozens of world-famous scientists from the Institute, who traveled around the world and clearly understood how the capitalist-world economy with its mostly contractual form of science financing works, quickly adapted to the changing situation and ensured a smooth transition. Now, when everybody understands everything and gives 'valuable' advice to others, much seems trivial. But then, it was necessary to find an effective way out of the situation and to take the right decision, which now produces significant results.

Now, a few words about Prokhorov as an educator: In fact, A.M. Prokhorov was an outstanding educator of young and not so young talents. But what rules did he follow? He followed a Democratic approach to everything and a sense of fairness and equity to everybody. Even his son, who now works at the Institute, often caught it from his father. No privileges! Everybody at the Institute knew that he would be listened to and supported. One cannot carve the future on regalia and merits of the past, which is why we had to prove our rightness every day, because the fact that you were right yesterday did not matter. Everyday someone was wrong in a dispute, but that was not a reason to mock. Tomorrow things could change. Everybody knew that we need to work and everything will be fine. Prokhorov's casual question: "What's new?" – And, almost at once, his answer with a smile on his face – "Nothing!" It was a usual, very conventional phrase to start a conversation the next day. Yes, last night we all went home, and this morning there could and should be news of a scientific nature, of course. The Institute is the Shrine of Science, where research is a continuous process.

In our lives we spend too much time in the laboratory, often losing touch with reality and with small things, of which our life outside the laboratory, in general, consists. We often need to do something for our child, to help our mother or to give a supporting hand to a close relative, etc. At the same time, there are very serious situations when help is most needed. And here (moreover, it was well known in the scientific world) the best solution was to ask A.M. Prokhorov. His heart will be open for you. Not only colleagues from our Institute but also from other Institutes asked him for help. And everybody knew that he will not refuse. And everybody knew that if there is an opportunity to help, he will certainly do it. I do not know whether time will come when it is no longer necessary to help people, but I know for sure that Prokhorov's office is not spacious enough to accommodate all those whom he helped.

Another distinguishing feature of A.M. Prokhorov was his natural ease with communication. He always showed patience and respect, no matter if his interlocutor was a student or a government bureaucrat. Communication with him was inspiring. Of importance was intelligence – intrinsic essence of civilization development – of the interlocutor. No wonder that in such situations people felt at ease, finding new opportunities for self-expression and having a surge of creative forces, which they really liked themselves.

I remember an interesting meeting of A.M. Prokhorov with Mr. M. Shikaya, the Tokyo Metropolitan governor, in Japan. I was fortunate to be part of this meeting. The Japanese, who already had a feeling about visitors from Russia, instructed our delegation for a few minutes about things they considered appropriate for their century-old principles and rules of good manners in their homeland. We found out that we could talk about flowers, nature and health. All other topics of conversation could be construed as improper. You should have seen the faces of these instructors a few minutes after a conversation between Prokhorov and Shikaya started. They talked as if they had known each other since childhood and they were incredibly happy with the possibility of communicating with each other. In this life, when eternity is born in the hands of such people, to spend time talking about flowers and bows just means not to respect each other. Apparently, this protective form of communication is introduced in Japan in the event of visitors from Russia, who can talk only about credits and dividends on favorable terms, which, of course, is extremely important today. This is humanly understandable, though still not to all.

A.M. Prokhorov was not only a scientist by profession. He was, as they say, a physicist to the bone. Moreover, his habits were even physically correct. Here is one of them: He loved it when it is warm, even hot (so hot that I would compare it with the Sahara Desert) in his office. "Why to warm the office with the warmth of human bodies? What is the average temperature of a normal person??? 36.6°C! So, here we are!" It was almost physically impossible to linger in his office for a long time. Heaters stood at the window, directly behind the visitor and quite close to the visitor's back. Thus, for someone it was thermodynamic equilibrium, but for someone it was nothing but thermal screening of the boss. However, it was only the smallness of the space surrounding him, because no space could accommodate the greatness of this man.

Now, there are a few facts about the great intuition of this great scientist. I will not talk about his anticipation of the laser era. Much has already been said about it and much more will be written by those who were next to him at that crucial period of his life. I should talk about things I witnessed myself and about projects I directly participated in.

It is difficult to overestimate the importance of lasers in solving the problems of medicine and biology. Even at the dawn of the laser revolution, when military men were overexcited about possible applications of lasers in the country's defense, A.M. Prokhorov began to actively introduce into the consciousness of the Institute staff and civilians the ideas about the effective use of laser methods of treatment of patients and the use of lasers in biological research. Today, in our country and abroad, well-known are the methods of laser treatment of urolithiasis, laser eye surgery, bloodless laser surgery, TB treatment, diagnosis and treatment of skin diseases, laser hair removal, etc. Now it is impossible to imagine how doctors could treat patients without lasers and laser methods of diagnostics and therapy.

One more example directly concerns military applications. Lasers can be used and actively applied in solving military tasks, and this fact is no longer a secret. They cut, melt, reduce the mechanical stability of structures, allow the transfer of mechanical momentum and provide a power regime of destruction of military equipment. Moreover, lasers can do all these things at considerable distances. No wonder that the attention of professionals from the Ministry of Defense was drawn to the prospect of the use of lasers for military purposes. To solve the problem of ballistic missile warhead interception in the framework of the Strategic Defense Initiative, our country and the USA spent a huge amount of money. To a large extent, this stage of confrontation with the United States undermined our economy and led to the collapse of the Soviet Union. But there is another regime of destruction of military equipment, i.e., functional impact regime (the Americans call it "smart" impact), which is, in fact, today the main one in research and applications. There is no need to "cut" or "drill holes", which requires a lot of energy and is totally ineffective. All you need is the timely laser impact on a target, which in a large number of cases will be sufficient to disrupt the mission. Prokhorov wrote about it in his letter to the Minister of Defense of the USSR. And just imagine, it was in 1973, i.e., 11 years before the beginning of Reagan's "Strategic Defense Initiative". If only the society were ready for it, if only ... On the contrary, our country got involved in this senseless race and lost a lot. Americans also lost. But they were much richer and it saved them.

Now about Prokhorov's fame in the world and his role in the development of mankind and in pushing society forward, which are the main goals of geniuses, who are born, once in a while, on the planet Earth.

The Nobel Prize is a recognized indicator of outstanding abilities of a particular individual. But we all well know that "not all yogurts are created equal". Among the several hundred Nobel laureates there are geniuses of mankind, who received this award for the revolutionary transformation of life on the Earth. The Nobel Prize for laser and maser principles of generation and amplification of electromagnetic radiation with the effect of stimulated emission in quantum transitions of atomic and molecular systems is one of them. Today, it is impossible to imagine our life without lasers and their wide range of applications. This discovery stands in one line with the electron's discovery, electromagnetism, nuclear energy, penicillin, evolutionary principles of biological life on earth, chemical transformations of elements, transistors (basic elements of electronics, computers), etc. Nevertheless, it is simply impossible to work out a kind of a unified scale of values of certain discoveries. Moreover, it is difficult to take on the function of an arbitrator in this competition. Here I want to resort to the help of the Internet – a highly respected tool in the hands of the scientific world. We will google "who is who" in the world, at least on the basis of a small sample of widely cited scientists, artists and politicians. On April 4, 2003, this list of several prominent figures with a coefficient of "scientific fertility" looks like this:

1. Aleksandrov A.P. – 108
2. Basov N.G. – 700
3. Brezhnev L.I. – 611
4. Gorbachev M.S. – 548
5. Yeltsin B.N. – 19
6. Keldysh M.V. – 715
7. Primakov E.M. – 151
8. Prokhorov A.M. – 2160!!!
9. Pugacheva A.B. – 8
10. Stanislavsky K.S. – 159
11. Townes Ch. – 1190
12. Chubais A.B. – 100

This is a certain projection of viewing the problem, and nothing more, but rather indicative and informative I should say. Having looked through all the pages related to one or another known name on the Internet, you become fully aware of the role of the individual in history and the range of his creative interests. But do not overestimate the role of this kind of projection. Evaluation, of course, must be multi-faceted and it is the business of historians. Nevertheless, I believe that sooner or later the situation will improve, and the role of science and its geniuses will occupy its rightful place in our country, where even today the hierarchy of values is still unfortunately significantly oriented towards those who squander public property, natural resources and give practically nothing in return to their Motherland.

A talented person is talented in everything! It is well known that "we should step back for better observation." As time goes by, we go further from the point of parting with our teacher and friend. The sharp pain of loss softens, and small, insignificant details, which are totally unimportant in the image of this giant of humanity, disappear. What is left is an increasingly growing, irresistible sense of the continuing impact of this GREAT MAN on us and on the development of physical science, to which we – his disciples and associates – continue to faithfully serve today. And yet, the soul overflows with gratitude to Fate for being for having years with this rock of a man, an outstanding naturalist of the 20th century and a patriot of Russia – Academician Aleksandr Mikhailovich Prokhorov.

Moscow, December 2001. Two days before New Year.

Chapter 2

HIGH POWER/ENERGY HIGH REPETITION RATE WIDE-APERTURE REPETITIVELY PULSED LASERS

INTRODUCTION

A technique for obtaining the repetitively pulsed operating regime in high-power/energy wide-aperture lasers is proposed and experimentally realized. In this regime, the laser emits a train of pulses with duration of 100–150 ns and a pulse repetition rate of several tens of kilohertz. The main properties of the pulsed regime are theoretically analyzed and the proposed technique is tested in detail employing a test-bench gas-dynamic laser. The results of the test confirmed the conclusions of the theoretical analysis. The possibility of a repetitively pulsed regime in high-power wide-aperture lasers realization without significant reduction in the average output power is experimentally demonstrated.

Currently, interest is increasing in high-power/energy lasers (>100 kW) employed in the solution of a variety of research and production problems. The existing sources of high-power radiation operate only in the CW or quasi-continuous low-frequency (below 300 Hz) repetitively pulsed regime with a long pulse duration (tens of microseconds). The development of lasers operating in a high-frequency (10–100 kHz) pulsed regime with a short pulse duration (100–150 ns) or the conversion of the existing lasers to this regime will considerably extend the field of application of high-power lasers, improve the efficiency of their use by factors of several tens, and enable the realization of qualitatively new effects [1]. For example, to eliminate the plasma screening effect in the radiation-material interaction process, weaken the thermal radiation defocusing in long paths, improve the energy extraction efficiency in wide-aperture lasers, etc.

At a high output power exceeding several kilowatts, however, organizing transient lasing modes based on high-frequency resonator modulation runs into several problems, which are caused by wide apertures of resonator elements and accordingly of the laser beam as well as by the high power density.

Presently known devices intended for resonator loss modulation may be conventionally divided into several classes: opto-mechanical, acousto-optic, electro-optical, and self-bleaching. In high-power lasers, only opto-mechanical devices can be used, which include transparent or reflective apertures. The remaining modulator types involve transmission optical elements.

In the ten-micrometer range of wavelength, all optical materials possess a relatively high (up to several percent) absorption coefficient, which is responsible for significant heat release and, in the long run, a fast degradation of these elements. The use of intracavity disc modulators in high-power industrial lasers is restricted by the output power of several hundred watts: due to the high power density inside the resonator, plasma is produced at the modulator aperture edges to cause modulator degradation or beam screening. In particular, the output power of the CO, laser investigated in Ref. [2] (a CW output power of 5 kW) lowered by two orders of magnitude when the laser was converted to the repetitively pulsed regime with the aid of a mechanical tull-aperture modulator. The approach proposed in Ref. [3] appears to be more promising – modulating the gain of the active medium rather than the cavity loss. In this work, the gain of the active medium was modulated by imposing a strong external pulsed magnetic field. However, in this case there emerged almost insuperable difficulties related to setup scaling for larger volumes of the active medium as well as to increase the modulation frequency and the pulse contrast ratio. The authors of Ref. [3] were able to raise the modulation frequency to only 10 Hz in a series of several hundred pulses. This resulted in a reduction in the output power in the repetitively pulsed regime by almost an order of magnitude compared to the CW regime. In the injection of external signal, the methods of modulating the gain of the active medium appear to be the methods of choice [4].

Our work is concerned with a new technique of modulation of the gain of the active medium by radiation self-injection. This technique can be applied to obtain a repetitively pulsed regime in the range of average output power of the order of 100 kW. The aim of our work is to theoretically substantiate and experimentally realize the repetitively pulsed regime of gas-dynamic CO_2 and Nd YAG lasers.

SUBSTANTIATION OF RESONATOR DESIGN

In the long wavelength lasers with a high average output power, unstable resonator configurations are commonly used because of a large cross section of the active medium. In resonators of this type, externally injected low-power beams may exert a significant effect on the characteristics of output radiation [4, 5].

One way to realize the control regime is the self-injection of radiation – extraction from the resonator and return of a part of radiation after changing its spatio-temporal characteristics. The transition to the transient lasing mode is effected through the modulation of the self-injecting beam. Earlier, a study was made of laser versions with radiation self-injection into the paraxial resonator region [4]. However, analysis showed that the power of the beam injected into the paraxial beam region should be comparable with the output laser power to efficiently control the resonator of a continuously pumped laser, unlike pulsed systems with regenerative amplification [6].

The self-injection of a part of output radiation through the resonator periphery is more efficient: on return to the paraxial resonator region, the injection power significantly rises due to the large number of passages to play the dominant part in the formation of output radiation.

The role of peripheral radiation was first investigated in Ref. [4]. In the case of a traditional resonator, the role of *waves converging to the resonator axis was found to be insignificant, because their source is a narrow region with a small relative area at the edge of the output mirror; accordingly, the power of the control wave injected into the resonator is low. This wave has a large divergence, and only its small part (of the* order of 1/*Nf,* where *Nf*>>1 is *the* Fresnel number) participates in lasing.

The effect of the injection wave on the resonator characteristics can be enhanced by matching the beam phase with the resonator configuration and increasing the radiation power returned. In this case, the propagation direction and the wave front curvature of the injection beam should be so matched with the resonator configuration that the injection beam concentrates, after a relatively large number of passages through the resonator, near the optical resonator axis and transforms to a divergent wave that forms the output radiation. The injection beam energy should be high enough to exceed, after its arrival to the resonator axis, the saturation energy of the active medium. The experimental data of the investigation of the effect of this kind of self-injection on lasing in the stationary mode were reported in Ref. [5].

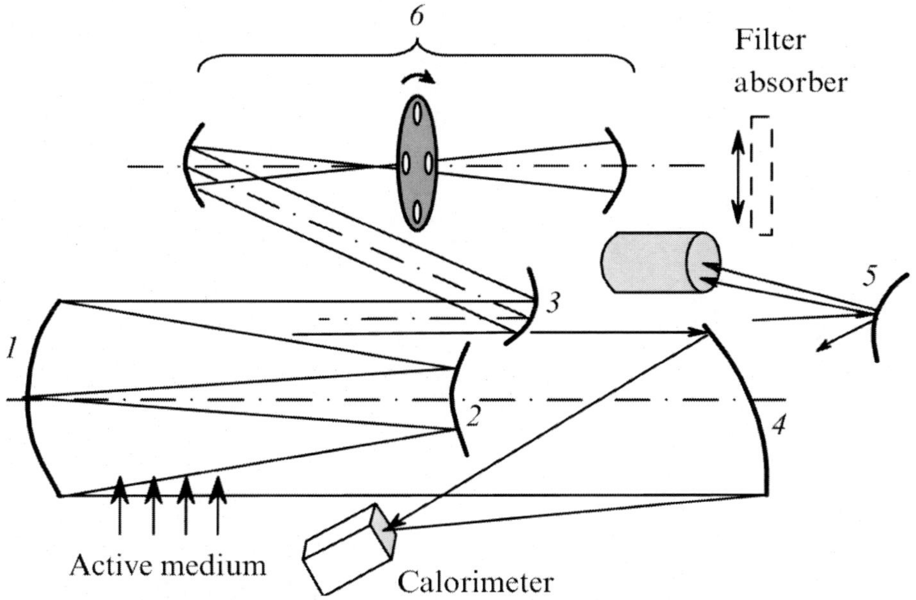

Figure 1. Scheme of the experimental setup: (*1, 2*) mirrors of the unstable resonator; (*3*) mirror coupler; (*4*) rotatory mirror; (*5*) deflecting mirror; (*6*) system for the formation of an injected beam.

The schematic of the setup, which realizes the repetitively pulsed radiation self-injection, is shown in Figure 1. The radiation is extracted from the resonator past the edges of a mirror (2), the mirror coupler (3) directs a part of the output beam to a system intended for the injection beam formation (6)*;* the beam is processed in the system (6), where it is modulated in power and acquires the requisite phase distribution, and is returned to the resonator with the aid of the same mirror. The above configuration was realized in a gas-dynamic CO_2 laser with the following parameters; the length of the active medium L_a = 1.2 m, the unsaturated gain g_0 =0.6 m^{-1}, the time it takes the active medium to transit the resonator $\tau = 0.92\times10^{-4}$ s, the relaxation time $\tau_r = 2.76\times10^{-4}$ s, the total round-trip time in the resonator $\tau_c = 4.2\times10^{-9}$ s, the luminescence lifetime $\tau_{\text{lum}} = 5$ s, the resonator magnification factor M = 1.45, and the diameter of output laser aperture a = 0.08 m.

The laser resonator is made of two spherical mirrors with rectangular apertures, which provided a geometrical amplification factor of 1.45. The active medium travels across the optical resonator axis. All theoretical and experimental data are given below for a laser with the above parameters. Such a temporal structure of radiation in the case of quazicontinuous mode of operation for pulsed lasers has a name – "rough pulse structure".

THEORETICAL LASER MODEL AND RESULTS OF NUMERICAL AND EXPERIMENTAL INVESTIGATIONS

For the initial theoretical treatment of lasing in a gas-dynamic laser with an unstable resonator and transmittance modulation, we will use the modified system of balance equations [7]. In the derivation of equations, the gain of the active medium was spatially averaged over the lasing volume. The gain was assumed to uniformly saturate, decreasing from its peak value (at the point of entry of the active medium into the resonator) to some minimal nonzero value (at the exit from the resonator) with the lateral coordinate. The resultant equations are written in the form coinciding with the form of equations in the case of quasi-stationary lasing mode:

$$\frac{dK}{dt} = \frac{2K}{\tau}\ln\frac{K_0}{K} - \frac{K[1+I+(\tau_r/\tau_{\text{lum}})K]}{\tau_r},$$

$$\frac{dI}{dt} = \frac{I}{\tau_c}(K-\delta) + \eta\frac{\tau_r}{\tau_{\text{lum}}\tau_c}K, \tag{1}$$

where $K_0 = 2L_a g_0$ is the averaged unsaturated gain-length product calculated in tracing around the resonator; $K = 2L_a g$ is the averaged saturated gain-length product in tracing around the resonator; $I = J/J_s$; J is the volume-averaged intensity; J_s is the saturation intensity; t is the current time; $\delta = \delta_0(1+\Delta(\nu))$ are the losses per round trip; $\delta_0 = -\ln(|\gamma^2|)$; $\Delta(\nu)$ is the modulating function; ν is the modulation frequency; and η is the fraction of spontaneous radiation power that remains inside the resonator after tracing around the resonator.

The first equation (1) is the equation of vibrational kinetics [8] of a pre-excited one-component (the lower working levels is not populated) active medium of a gas-dynamic CO_2 laser. The second equation describes the formation of radiation in the propagation through the resonator. The characteristics of active medium and radiation are averaged over the volume, and that is why the equations do not contain directional derivatives and depend only on time.

To determine the conditions ensuring the repetitively pulsed operating regime, the system of equations (1) was considered in the perturbation-theory approximation relative to the small parameters

$$\frac{\Delta I}{I_s}, \frac{\Delta K}{\delta_0}, \frac{\Delta \delta}{\delta_0},$$

where ΔI is the amplitude of the deviation of output radiation intensity from the stationary value; $I_s = 2(\tau_r / \tau) \times \ln(K_0 / \delta_0) - 1$ is the normalized output radiation intensity for the stationary lasing; ΔK is amplitude of the deviation of the gain-length product from the stationary value; $\Delta\delta$ is the transparency modulation amplitude; $\delta = \delta_0 + \Delta\delta \cos \omega t$; and ω is the circular frequency. The last-named quantity is related to the above-introduced modulation frequency ν in the usual way: $\omega = 2\pi\nu$.

In this approximation,

$$\frac{\Delta I}{\Delta I_s} = \left(\frac{\omega_{\text{res}}\tau}{2}\right)^2 \left\{\frac{(\omega\tau/2)^2 + 1}{(\omega\tau/2)^2 + [(\omega\tau/2)^2 - (\omega_{\text{res}}\tau/2)^2]^2}\right\}^{1/2}, \tag{2}$$

$$\frac{\Delta I}{\Delta I_s} = \frac{\Delta\delta\tau}{\tau_c}\left\{\frac{(\omega\tau/2)^2 + 1}{(\omega\tau/2)^2 + [(\omega\tau/2)^2 - (\omega_{\text{res}}\tau/2)^2]^2}\right\}^{1/2}, \tag{3}$$

where $\Delta I_s = (\Delta\delta / \delta_0)(\tau_r / \tau)$ are the quasi-stationary intensity fluctuations (for $\omega \to 0$), and $\omega_{\text{res}} \approx (I_s\delta_0/\tau_c\tau_r)^{1/2}$ is the resonance circular frequency.

The transition to the repetitively pulsed regime necessitates the fulfillment of two conditions: (i) the transmittance fluctuations should be fast enough, because otherwise the output radiation power will vary in the quasi-stationary manner; (ii) the value of ΔI should be large enough for the radiation intensity to be modulated to a near-zero value.

The former condition is satisfied for $\nu \geq 2/\tau$ and the second for

$$\frac{\Delta I}{I_s} \geq 1 \to \Delta\delta \geq \frac{\tau_c}{\tau}\left\{\frac{(\omega\tau)^2 + [(\omega\tau)^2 - (\omega_{\text{res}}\tau)^2]^2}{(\omega\tau)^2 + 1}\right\}^{1/2}.$$

For the laser investigated, $\nu_{\text{res}} \approx 100$ kHz, and the repetitively pulsed regime is realized for $\nu > 20$ kHz, $\Delta\delta / \delta_0 > 0.02.$

The resonance field can be represented as a super-position of two waves – the ordinary divergent wave and the convergent one, which transforms to a divergent wave in the incoherent summation in the paraxial resonator region. The transparency δ of the resonator with laser-radiation self-injection, taking into account the diffraction transformation of the convergent wave to the divergent wave in the paraxial resonator region, is defined by the relationships:

$$\delta = 1 - |\gamma^2|, \quad |\gamma^2| = \frac{1}{M^2} + \frac{s}{|\gamma^{4N}|}, \tag{4}$$

where

$$N \sim \ln\left[\frac{a^2}{\lambda L_r}\left(1 - \frac{1}{M}\right) + 1\right] \Big/ 2\ln M$$

is the number of transits required of the beam injected into the resonator to find itself in the region of paraxial diffraction transformation; $s = S/\pi a^2$ is the relative injection beam area; S is the injection beam area; and λ is the radiation wavelength.

Figure 2 shows the calculated resonator transmittance against beam fraction returned to the resonator. One can see that the modulation amplitude of resonator losses amounts to 30%-50% of the losses of the basis resonator (without self-injection) when the power of the beam returned to the resonator is about 5% of the output beam power. This loss modulation amplitude is sufficient to ensure the repetitively pulsed operating regime.

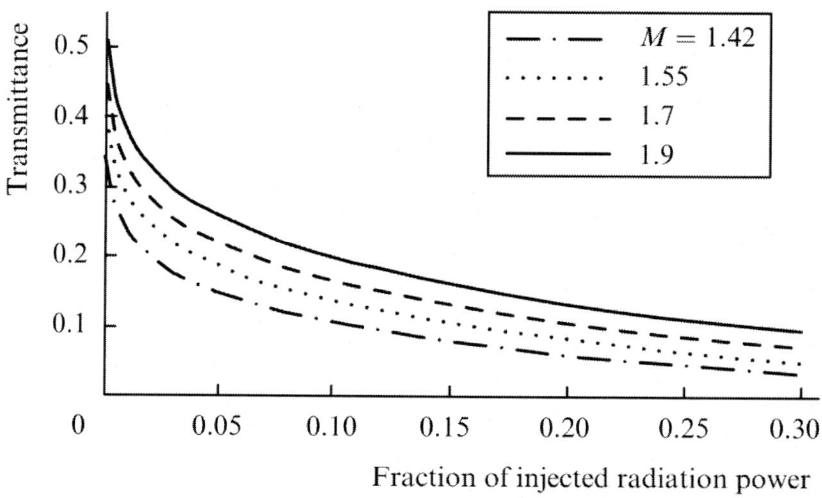

Figure 2. Dependence of laser resonator transmittance on the fraction of radiation power injected into the resonator for different values of resonator magnification factor *M*.

To derive qualitative estimates of laser operating modes with self-injection, we considered the energy and time characteristics of the laser with the parameters specified above. The system of equations (2) was numerically investigated employing the Runge-Kutta method. Figure 3 gives the time dependences of the output power for several values of the modulation frequency and depth; the geometrical resonator amplification factor is M = 1.45. The calculated data are in qualitative agreement with the notions of the dynamics of quantum processes occurring in lasers [8].

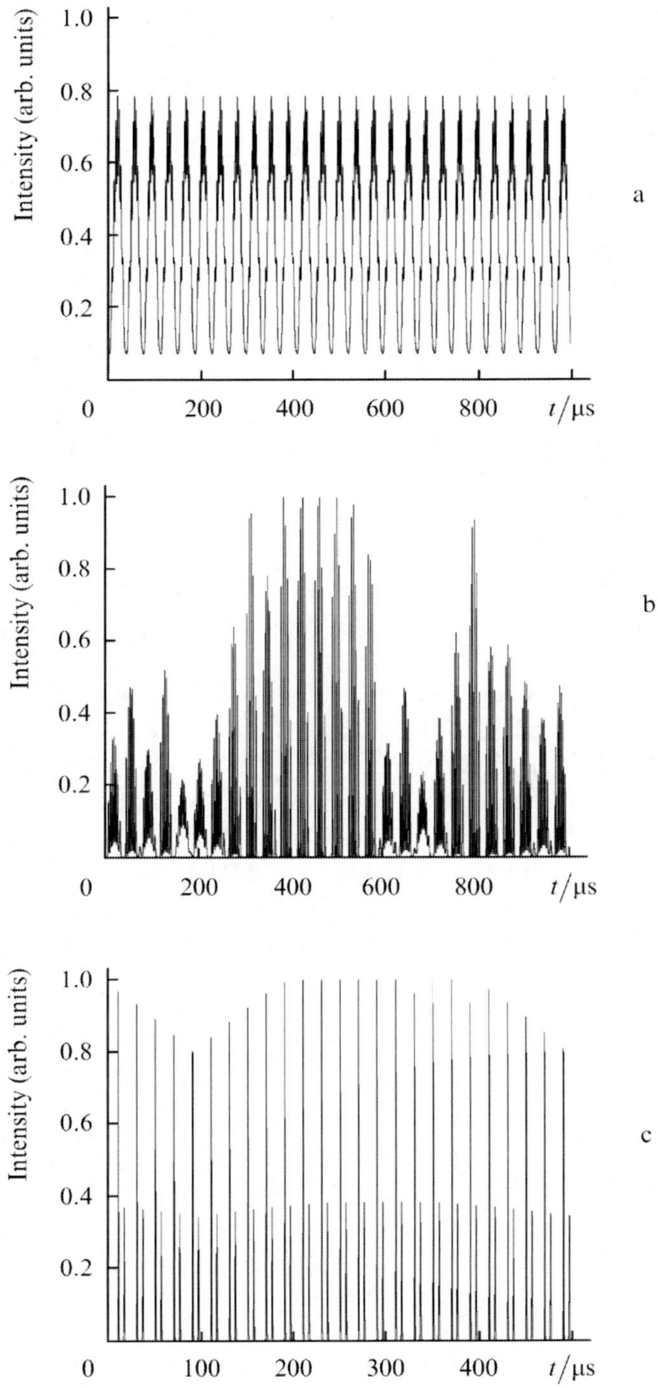

Figure 3. Rough pulse temporal structures of output laser radiation for a modulation depth of 5% and a modulation frequency of 27 kHz (a), 5.8% and 27 kHz (b), and 5% and 50 kHz (c); $M = 1.45$.

Numerical calculations indicate that the pulses of output radiation power reproduce the modulation pulses in shape and duration for modulation frequencies up to 20-25 kHz (Figure 3a). When the transmittance modulation depth is raised above the critical value, within the modulation pulse length there emerge separate power peaks, whose total number (4-8) is close to the resonance-to-modulation frequency ratio. Their modulation depth amounts to 100% (Figure 3b). For modulation frequencies $2/\tau < \nu < \nu_{res}$, the laser goes over to a mode close to the *Q*-switching mode (Figure 3c). In this case, there occurs not only an increase in the modulation depth of output radiation power, but a change in the characteristic pulse structure – the envelopes of individual pulses and their intern till peak structure become more regular. The peak intensity in this mode exceeds the stationary intensity by more than a factor of 10. In the CO_2 laser case, the duration of an individual peak of the structure is comparable with the pulse duration of a free-running pulsed CO_2 laser (hundreds of nanoseconds).

Therefore, when the modulation frequency is lower than ν_{res}, the lasing exhibits two characteristics oscillation constituents – the low-frequency oscillation, defined by the modulation frequency, and the high-frequency oscillation, defined by the eigenmodes of the resonator-active medium system, whose frequency is close to ν_{res}. When the modulation frequency is made greater than ν_{res}, the forced oscillations manifest themselves in the form of the high-frequency component, while the slow oscillations are the natural oscillations of the system at the resonance frequency. The results of numerical calculations are indicative of the feasibility of the repetitively pulsed regime in wide-aperture lasers described by the model (1).

The results of numerical calculations were experimentally verified on a test-bench CO_2 gas-dynamic laser whose parameters were given in the foregoing. For a fuel, use was made of carbon monoxide (CO), with air as the oxidizer. The typical output power was equal to 50 kW. To preclude the damage to optical elements of the laser, the output power was lowered by lowering the flow rate of the working components. When the laser was operated in the CW mode, the output power was equal to about 10 kW. Since the dements of the test-bench structure were not cooled, the duration of rims was limited by the heat capacity of resonator elements and combustion chamber and was equal to 3 s, the nominal power settling time was 0.3 s from the onset of mixture combustion.

The optical configuration of the experimental setup was similar to that diagrammed in Figure 1. A part of the output laser radiation (about 20%) was diverted by an inclined metallic mirror to the injection beam formation system consisting of two spherical mirrors with conjugate focal planes. In the vicinity of the focal plane there formed the waist of the branched part of the laser beam, and a modulator was placed near the waist. The modulator location was selected so that the laser beam completely filled the aperture of every round hole in the modulator disk. The modulator was a rotating metal disk with openings along its perimeter. In experiments, use was made of disks with 150 and 200 drilled holes with respective diameters of 4 and 2 mm and a filling factor (the ratio between the open state duration and the total period) of 1:2. The maximal modulation frequency was equal to 33 kHz.

A VIGO SYSTEM PD-10.6-3 photodetector was used as a radiation detector of the measuring system. The detector enabled measuring both the temporal structure of the signal and its constant component. The output beam is characterized by a high power density, which is many times higher than the optical breakdown threshold and upper bound of dynamic range of the detector, the radiation was attenuated employing the geometrical factor fa convex deflecting mirror (5)] and optical attenuation filters. The signal generated by the photodetector entered a pre-amplifier. After the preamplifier and a cable line, the signal was recorded by a broadband digital storage Tektronix THS710 oscilloscope. The transmission band of the path was limited primarily by the preamplifier and was equal to ~50 MHz.

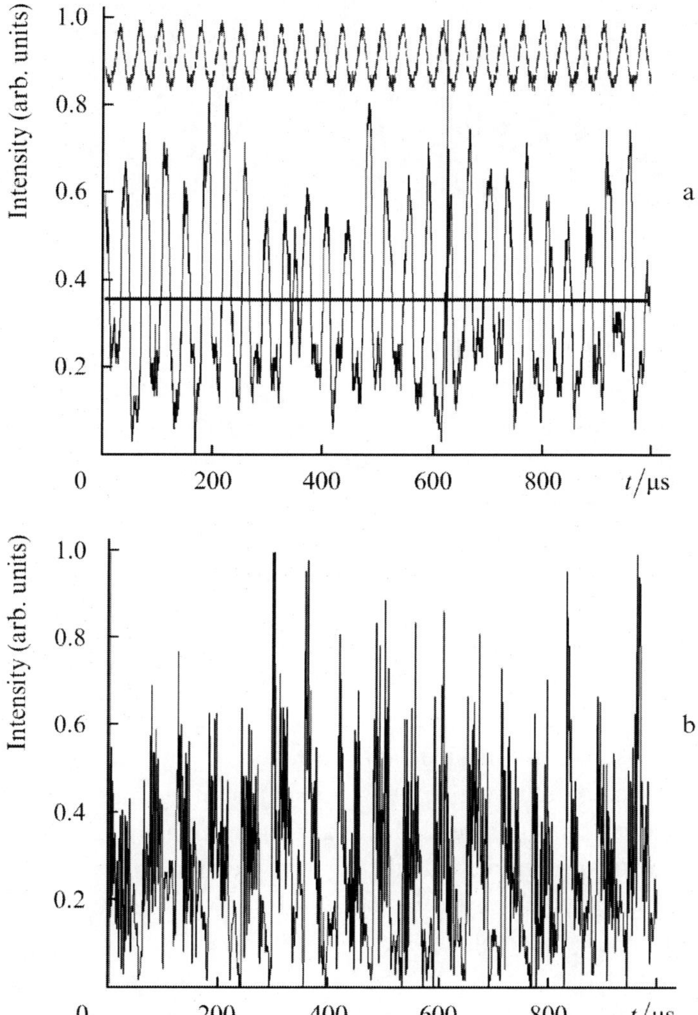

Figure 4. Rough pulse temporal structures of laser radiation for a modulation depth of ~ 3% and a modulation frequency of ~ 27 kHz (top – modulating signal, bottom – output laser signal) (a) and for a modulation depth of 7% and a modulation frequency of ~ 25 kHz (b).

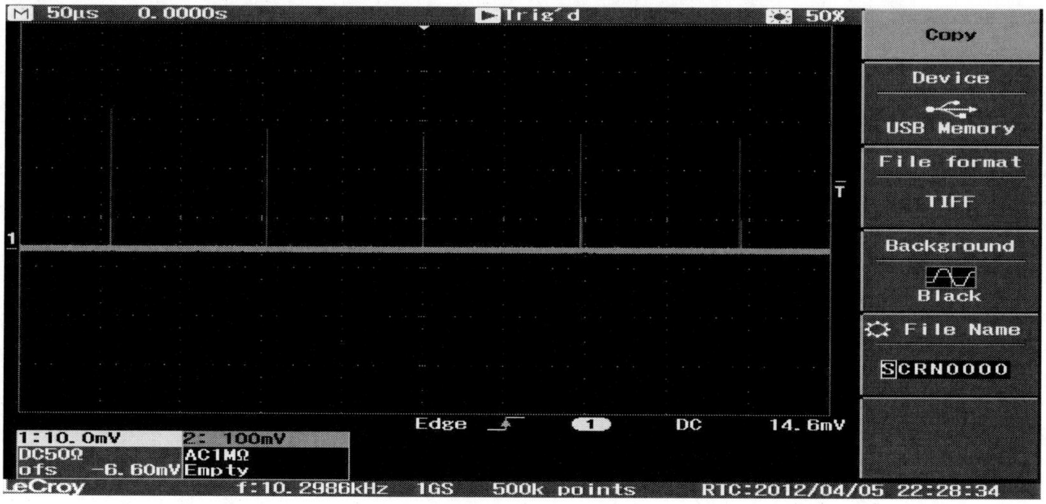

Figure 5. High repetition rate P-P mode of operation for high power Nd YAG laser.

The average output power was measured with a calorimeter cooled by running water. A mirror 4 focused the radiation onto the calorimeter. In the CW mode, the constant-level signal was recorded with a noise component, which did not exceed 5% of the constant signal level. The oscilloscope traces of laser output in the case of radiation modulation are depicted in Figure 4. For a modulation frequency of about 27 kHz and a modulation depth of 2%−3%, the quasi-stationary modulation regime is realized (Figure 4a). In this case, the laser radiation exhibits intensity fluctuations consistent with the modulating signal, with the output power departing from the average value by a factor of three. This regime agrees well with the operating regime shown in Figure 3a. When the modulation depth was increased to 7%−8%, the laser passed to the repetitively pulsed operating regime (Figure 4b). In this case, lasing took place in the form of a train of 5−10 pulses within one cycle of the open modulator state. The duration of an individual pulse was about 200 ns. We emphasize that the recorded pulse duration was limited by the measuring path bandwidth, which was equal, as noted above, to 50 MHz. The amplitudes of individual pulses exceeded the average value by factors of 6.5−11. This regime agrees well with the regime calculated by expressions (1) and presented in Figure 3b. Note that the average output power in the repetitively pulsed regime was equal to the output power in the CW laser operating mode.

The technical characteristics of the modulator in use (the maximal modulation frequency was 33 kHz) did not permit the high-frequency modulation modes realization presented in Figure 3c. Good agreement between the experimental and theoretical data for frequencies ranging up to 25−30 kHz testifies to the adequacy of the proposed model and the possibility of employing this method at higher frequencies to convert a CW laser operation to the operating mode with regular pulses structure. Round trip time of the resonator geometry and repetitively pulsed regime frequency of pulses matching should be taken into account. The efficient method of regular structure for P-P operating regime for CO_2 and LD pumped Nd YAG lasers has been realized. The studies of the power and temporal characteristics of the laser radiation show that the developed lasers have a very high efficiency of energy extraction close to that of a CW laser mode of operation. The pulse peak power to average power ratio of more than two orders of magnitude after transformation of CW into P-P mode of operation had been demonstrated. The prospects of efficient and compact P-P/CW laser systems are open.

CONCLUSION

We have demonstrated both theoretically and experimentally, the feasibility of converting a high-power/energy wide-aperture CO_2 and Nd YAG lasers to the repetitively pulsed lasing regime by self-injecting the modulated fraction of output radiation without decreasing the output power compared to the stationary lasing. In this case, the peak output power can exceed the average output power by more than an order of magnitude. Repetitively pulsed modulation with a pulse length of 100 ns, a peak output power greater than 100 kW and an average output power coinciding with the power of stationary lasing (10 kW for CO_2 and 1 kW for Nd YAG) were experimentally obtained. The applicability of the proposed method of laser conversion to the repetitively pulsed lasing regime is limited by the threshold of optical breakdown on the modulator aperture only, which can be easily eliminated by decreasing the injected signal.

REFERENCES

[1] Apollonov V.V., Prokhorov A.M. et al. *Proc. SPIE Int. Soc. Opt. Eng.,* 3574, 2 (1999).

[2] Husmann A., Niessen M., Grumbel F., Kreutz E.W., Poprawe R. *Proc. SPIE Int. Soc. Opt. Eng.,* 3343, 759 (1998).

[3] Hager G.D., Anderson B., et al. *Proc. SPIE Int. Soc. Opt. Eng.,* 4065, 646 (2000).

[4] Anan'ev YuA, Optical Resonators and Laser Beams, Moscow: Nauka, (1990).

[5] Vagin Yu.S. *Tr. Fiz. Inst. Akad. Nauk SSSR,* 113, 115 (1979).

[6] Apollonov V.V., Alcock A.J. Baldis H.A., Corcum P.B., Taylor R.S. *Opt. Lett.*, **5**, 333 (1980).

[7] Breev V.V. et al. Energy Diagrams and Investigation of Characteristics of Fast-Flow Stationary CO_2 Lasers, Moscow: Izd. IAE im. I.V. Kurchatova, (1982).

[8] Khanin Ya.I. Dynamics of Quantum Oscillators, Moscow: Sov. Radio, (1975).

Chapter 3

ELECTRICAL DISCHARGE PULSE-PERIODIC LASERS

SELF-SUSTAINED VOLUME DISCHARGE FORMATION IN CO_2 LASERS

Introduction

A self-sustained volume discharge (SSVD) can be established in a gas by creating primary electrons, the density of which should exceed a certain minimum value n_{min} throughout the discharge gap [1−3]. Various methods for pre-ionization of a gas in the discharge gap have been developed for this purpose [4−7].

Primary electrons are created by these methods directly in the discharge gap, which sometimes causes difficulties in the establishment of conditions necessary for the formation of a SSVD. For example, high voltages are needed for the formation of an initial plasma when the pre-ionization source and the active medium are combined in the same volume [8], whereas pre-ionization with ultraviolet radiation may be ineffective because of strong absorption of such radiation in a medium [9], and if soft X rays are used, it is necessary to ensure rigid synchronization of the X-ray and pump sources [10].

In mixtures of gases typical of CO_2 lasers, losses of electrons due to trapping are relatively small at low values of $E = p$, where E is the electric field intensity and p is the gas pressure. The trapping coefficient is considerably less than the absorption coefficient of ultraviolet radiation for the same mixtures [10]. It follows that, in principle, it should be possible to create primary electrons in a density needed for the formation of a self-sustained volume discharge at a considerable distance from an ionization source and then transport the electrons to the gap by drift in an electric field.

A method of formation of a SSVD by filling the discharge gap with a flux of electrons drifting in an electric field, without preliminary ionization of the whole discharge volume, was proposed and implemented [11−13]. The source of electrons was plasma formed in an auxiliary discharge initiated under a grid cathode. This method also made it possible to establish a SSVD in a system of electrons with a strongly inhomogeneous electric field in the discharge gap.

Physical Model of SSVD Formation

The SSVD formed by filling the discharge gap with primary electrons was studied in the electrode system with a uniform initial distribution of the electric field (the anode with the Chang profile [20] – a flat mesh cathode) with a discharge gap measuring 6 to 10 cm in length. In these experiments use was made of the configuration presented in Figure 1. When U_0 was less than the static breakdown voltage of the discharge gap, U_{eT}, use was made of a scheme with pulsed application of the voltage exceeding U_{CT} to the discharge gap. In the latter case, the filling of the discharge volume with electrons from the plasma of the auxiliary discharge was performed either by applying a bias to the discharge gap [21], or by artificially increasing the duration of the leading edge of the voltage pulse applied to the discharge gap [21, 22]. The value of T_f was chosen so that the electrons can reach the anode before the onset of the ionization processes in the discharge gap.

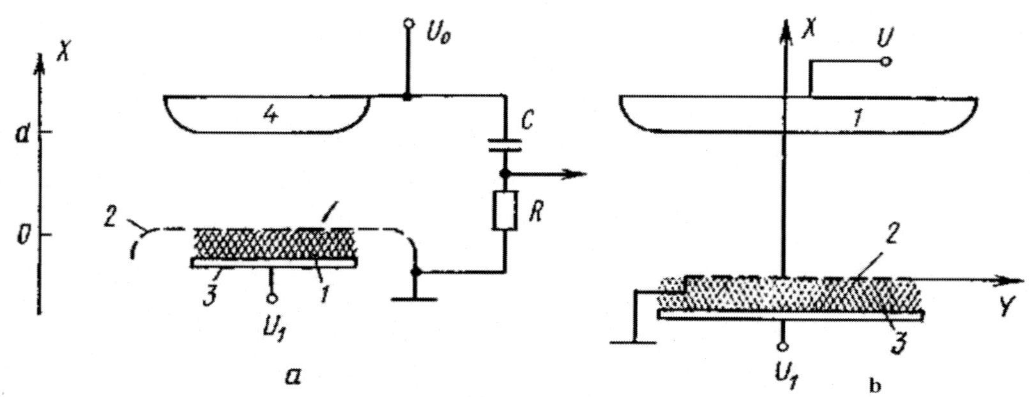

Figure 1. Discharge systems with profiled (a) and non-profiled (b) electrodes.

The conditions of the SSVD formation in initial highly non-uniform field distribution in the discharge gap were performed using the setup schematically shown in Figure 1b. The electrode system (d = 10 cm) was formed by a planar anode 1 (a rectangular plate measuring 18×50 cm in size, curved along a radius of 10 mm) and a cathode 2 with sharp edges (a rectangular brass mesh measuring 10×40 cm in size). The surface area covered by auxiliary discharge 3 under the mesh was slightly larger than the cathode area.

In order to reduce the working voltage [23, 24] and increase the duration of stable combustion of the SSVD [25, 26], easily ionizable substances such as tripropylamine and triethylamine at a vapor pressure of were added into the mixture.

In the experiments we controlled the lower limit for the SSVD ignition with respect to the voltage U_{min} (determined by the voltage of the pump source), and the SSVD stability limit with respect to the $CO_2 - N_2 -$ He gas mixture composition and the energy input. The SSVD uniformity was monitored visually by photos of the glow discharge, as well as by the small-signal gain and its distribution over the aperture of the amplifier.

The electron flux current in the discharge gap was deduced from the voltage drop across a resistor R. The charge carried by the flux was found from the change in the voltage across a capacitor C for R = 0. The contribution to the gap conductance made by the photoionization of the gas in the gap by the auxiliary-discharge radiation was estimated from the charge of the positive ions formed in the gap were carried out by applying a negative bias voltage U_0).

The following types of an auxiliary discharge were investigated:

1) A barrier discharge [14] in which a wire insulated with polyethylene was used as the electrode.
2) A corona discharge initiated by a multi-pin electrode, stabilized by limiting the current through the pins by a ballast resistor [15].
3) A volume discharge between a grid (discharge gap cathode) and a metal cathode, initiated as described in Ref. [16].
4) A volume discharge between a semiconductor cathode (carbon-filled rubber with a bulk resistivity [17] and a grid (discharge gap cathode) initiated without pre-ionization [18].
5) A multichannel discharge on the surface of a high-permittivity insulator limited by the insulator capacitance [19].
6) A multichannel discharge on the surface of a semiconductor (carbon-filled rubber [17]) limited by the semiconductor resistance (this auxiliary discharge was realized in the same way as a multichannel discharge on the surface of a ceramic described in Ref. [19]).

In all these auxiliary discharges, an electric field was applied to auxiliary discharge plasma. A special auxiliary electrode was not essential. The auxiliary electrode is the one used to apply the voltage pulse to ignite the auxiliary discharge.

The propagation of primary electrons in the discharge gap (one-dimensional case) in the absence of ionization multiplication is described by the system of equations and boundary conditions for the electric field $E(x, t)$, the density of electrons $n(x, t)$ and negative ions $N_i(x, t)$ [27]:

$$\partial n / \partial t + \partial (n\mu E) / \partial x = -\alpha n\mu \,|\, E \,|; \tag{1}$$

$$\partial N_i / \partial x = \alpha n\mu \,|\, E \,|; \tag{2}$$

$$\partial E / \partial x = -(N_i + n)e / \varepsilon_0; \tag{3}$$

$$\int_0^d E dx = -U_0, \tag{4}$$

$$n(0,t) = \text{const} = n(0); \tag{5}$$

$$N_i(x,0) = 0; \tag{6}$$

$$n(x,0) = 0 \ \text{при} \ x > 0, \tag{7}$$

where α, μ, and e are the attachment, mobility and charge of electrons, respectively; ε_0 is the dielectric constant. The X-axis is directed from the cathode ($x = 0$) to the anode

($x = d$); the electron density at the cathode surface $n(0)$ is assumed to be constant over time (the escape of the electrons from the cathode to the displacement field is compensated for by the influx from the plasma of the auxiliary discharge). In expressions (1)–(7) the mobility of the negative ions is negligible.

System (1)–(7) was solved using an explicit difference scheme [21]. To simplify the calculations we assumed that $\alpha = \text{const}$.

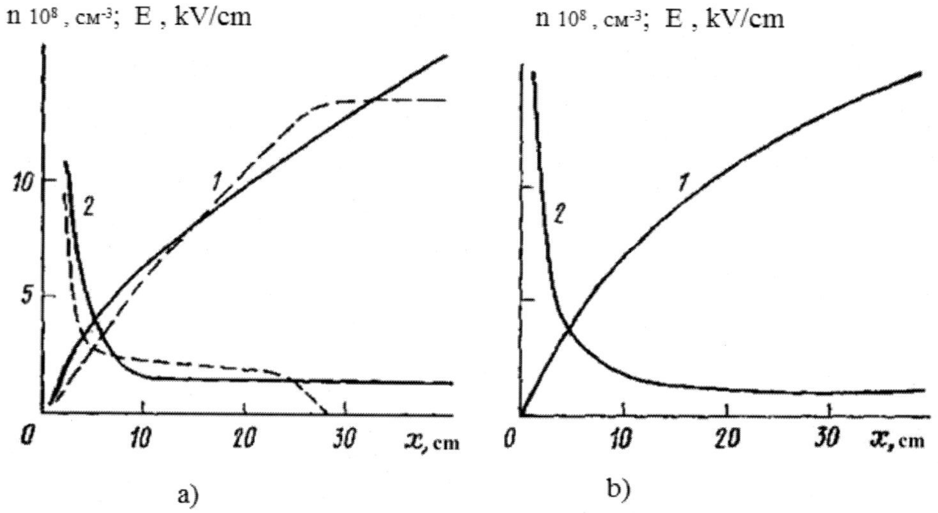

Figure 2. Calculated dependences $E(x)$ (curves denoted by 1) and $n(x)$ (curves denoted by 2): (a) $t = 2$ mks (dashed curves) and 4-6 mks (continuous curves); (b) $U = 400$ kV, $T_f = t = 8$ mks, $n(0) = 10^{10}$ cm^{-3}.

Figure 2a shows the dependences $n(x)$ and $E(x)$ for various time points at $U_0 = 400$ kV, $d = 40$ cm, $\mu = 10^3$ cm^2/Vs, $n(0) = 10^9$ см$^{-3}$. It is seen that in the absence of the electron loss through the attachment, stationary distributions $n(x)$ and $E(x)$ are established in the discharge gap for $t > t_0 = d/E_0\mu$, where $E_0 = U_0/d$.

The discharge gap accumulates a volume charge of negative ions, which prevents the propagation of primary electrons. Calculations performed for different a and d showed that at the time $t = t_0$, the electron densities near the anode $n_\alpha(d)$ (at $\alpha \neq 0$) and $n(d)$ (at $\alpha = 0$) are related as by the expression:

$$n_\alpha(d) \approx n(d)e^{-\alpha d}. \tag{8}$$

In conventional CO_2 lasers utilizing atmospheric pressure $CO_2 - N_2 - He$ gas mixtures, $E_0 \leq 3$ kV/cm is less than 10^{-2} см$^{-1}$ [11]. Therefore, when $d < 1$ m and $E_0 \geq 3$ kV/cm, the loss of electrons through the attachment does not significantly affect the propagation of primary electrons, and the analysis of the possibilities of the method can rely on the model of a stationary primary electron ($\alpha = 0$ and $t > t_0$).

Approximate steady-state solution to system (1)–(7) at large $n(0)$, when

$$\delta = 3\varepsilon_0 E_0 / 4en(0)d << 1 \qquad (9)$$

has the form [28]

$$E(x) = \frac{3}{2} E_0 \sqrt{\frac{x}{d} + \delta^2}; \qquad (10)$$

$$n(x) = \frac{3}{4} \frac{\varepsilon_0 E_0}{ed} \sqrt{\frac{d}{x + \delta^2 d}}; \qquad (11)$$

$$j = \frac{9}{8} \mu \varepsilon_0 \frac{E_0^2}{2} (1 - 3\delta^2), \qquad (12)$$

where j is the current density of primary electrons.

Let us perform estimations. Let E_0 = 1 kV/cm, d = 50 cm, then from (11) $n(d) \approx 10^7$ cm^{-3} clearly exceeds N_{min}, required for the SSVD formation [21]. In this case, condition (9) is fulfilled already at $n(0) \approx 10^8 - 10^9$ cm^{-3}. It will be shown below that such low densities of the electrons at the cathode can provide almost any type of auxiliary discharge.

Thus, our estimates confirm the possibility of obtaining the SSVD by filling the discharge gap with primary electrons, which are necessary for the SSVD formation process by realization of high concentrations of primary electrons at large distances from the cathode. Note some features of this method which follow from the considered model.

It is seen from (11) that at a certain distance from the cathode ($x >> \delta^2 d \approx 10^{-3} d$ при E_0 = 1 kV/cm, d = 40 cm, $n(0)$ = 10^9 cm^{-3}) under the condition of (9) $n(x)$ is independent of $n(0)$. Therefore, the discharge gap will be filled by electrons uniformly (in a plane perpendicular to x), even in the case of a significant non-uniformity of the $n(0)$ distribution over the cathode area. This makes it possible to produce a SSVD without small-scale inhomogeneities (typical of systems with UV pre-ionization) caused by shading of the discharge volume by a mesh cathode [29], and to increase the small-signal gain and efficiency of CO_2 lasers.

It follows from (10) that, despite the relatively low concentration of the primary electrons in the discharge volume, the presence of the uncompensated negative charge of primary electrons leads to a significant distortion of the electric field in the discharge gap. As a result, the ionization processes will develop in the first place at the anode, where the electric field can be enhanced by 1.5 times compared to E_0. At the same time, the development of ionization at the cathode is difficult, because the field is close to zero due to the screening of the cathode by the volume charge of primary electrons. Obviously, the effect of the cathode screening when filling the discharge gap with electrons should manifest itself in the real geometry of the discharge gap. Therefore, it is advisable to try to find the conditions for obtaining the SSVD in systems of non-profiled electrodes, which would greatly simplify the creation of electric discharge lasers.

Distortion of the electric field in the discharge gap by a volume charge of primary electrons should result in a voltage drop (this voltage is used to produce a SSVD), which is an obvious advantage of the proposed method. However, in a system with preliminary displacement U_0, which is used to fill the discharge gap with electrons, and subsequent to apply the voltage pulse with the amplitude U to the discharge gap, degree of enhancement of the electric field at the anode (with respect to the U/d) $K_y = 1 + U_0/2U$ may be small, because it is not always acceptable increase U_0 to values close to U_{st}. The degree of distortion of the electric field can be increased by using a scheme with an artificially prolonged voltage front [22]. Figure 2b shows the results of numerical solutions of equations (1)–(3), (5)–(7) for $n(x)$ and $E(x)$ at $\alpha = 0$ with the boundary condition

$$\int_0^a E dx = -\frac{U}{2}\left(1 - \cos\frac{\pi t}{\tau_\Phi}\right), \tag{13}$$

which corresponds to the law slew of the voltage rise in the discharge gap in this scheme. To completely fill the discharge gap with the electrons, it must also satisfy the condition $T_f = 2d^2 / \mu U$. It is seen from Figure 2b that when $t = T_f = 2d^2 / \mu U$, as in the case of constant displacement in the discharge gap, the gain $E(d)$ is reached, which U/d is 1.5 times higher relative to U/d at multiple reductions of $E(0)$.

EXPERIMENTS

Figures 3a and 3b show oscillograms of the electron charge q(t) extracted from the discharge gap and of the voltage U(t) between the gap cathode and the auxiliary-discharge electrode, recorded for a mixture of N_2 with tri-propylamine obtained for the case when $d = 2$ cm and the electrode area was $S = 200$ cm^2. The auxiliary discharge was a multichannel discharge on the surface of a ceramic (with the permittivity $\varepsilon = 2000$), limited by the capacitance of the ceramic itself [19]. The shape of $U(t)$ was set by the parameters of a special RC circuit. A comparison of the $q(t)$ oscillograms recorded for different polarities of the voltage $U(t)$ indicated that the main rise of $q(t)$ occurred in the region where $dU/dt < 0$. When a negative bias voltage was applied to the discharge gap, the form of the $q(t)$ oscillograms was independent of the sign of dU/dt. In accounting for these results one should bear in mind that the sign of the electric field in the gap between the ceramic surface and the discharge gap cathode was the same as the sign of dU/dt. Therefore, for $dU/dt < 0$ some of the electrons were expelled from the auxiliary-discharge plasma into the discharge gap across the grid cathode and drifted toward the anode in the bias field ($U_0 > 0$). For $dU/dt > 0$ or $U_0 < 0$ it was not possible to drive electrons beyond the grid and in the discharge gap the flux of the current was then entirely due to the photoionization of the gas by the auxiliary-discharge radiation.

Figure 3c shows the oscillograms of the voltage $U(t)$ and of the electron current in the discharge gap $I(t)$ for an oscillatory form of the voltage $U(t)$. It is clear from these oscillograms that a considerable current in the discharge gap was recorded only for $dU/dt < 0$, irrespective of the sign of $U(t)$.

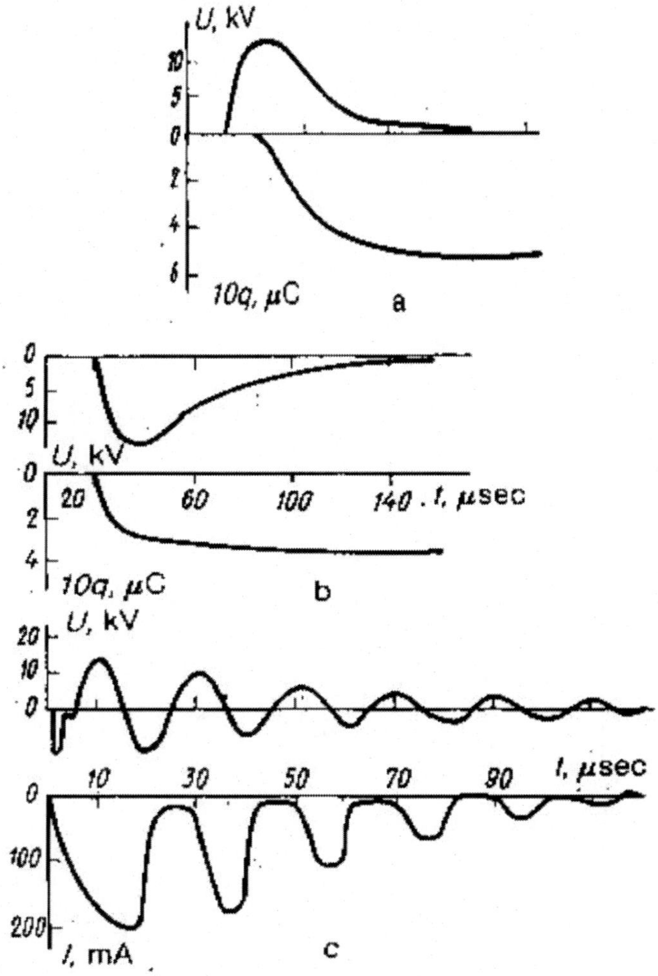

Figure 3. Oscillograms of the voltage $U(t)$ on the auxiliary electrode, of the electron charge $q(t)$, and of the electron flux current $I(t)$ in the case of positive (a) and negative (b) polarities, and also in the case of oscillatory nature of the voltage $U(t)$ (c).

It should be stressed that we investigated long-duration auxiliary discharges in order to identify the conditions for the formation of an electron flux from the auxiliary-discharge plasma itself. Under these conditions, in spite of the presence of an easily ionizable substance in the gas, the contribution to the discharge gap conductance of the electrons arriving from the auxiliary-discharge plasma was (for suitable signs of dU/dt and U_0) approximately ten times higher than the contribution made by the photoionization of the gas by the auxiliary-discharge radiation even when the latter was of the surface type. Under these conditions the contribution of the photoionization to the charge of the electrons crossing the gap did not exceed 10% of the contribution of the electron flux for a corona discharge and it was not greater than 1% in the case of barrier volume discharges (these values were obtained for $d = 6$ cm, $U_0 = 30$ kV, mixture of N_2 with tripropylamine, $P = 1$ atm) and the presence of an easily ionizable substance in the gas had practically no influence.

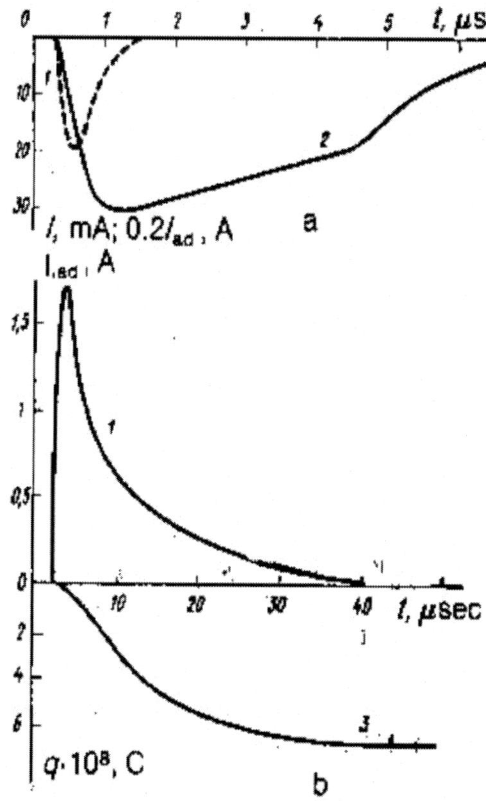

Figure 4. Oscillograms of the volume auxiliary discharge current (1) from metal (a) and semiconductor (b) electrodes, of the electron flux current (2), and of the charge carried by this flux (3).

When the auxiliary discharge pulse was shortened (by reducing the inductance or the active resistance in the circuit of this discharge), the contribution of the photoionization to the discharge-gap conductance increased and, in principle, it was possible to form an electron flux from a plasma created by the ultraviolet radiation of the auxiliary discharge in the cathode region (and also from a plasma on the surface of a plasma cathode [30] or a barrier discharge plasma in systems described in Ref. [16]).

As already pointed out, the homogeneity of a SSVD could be improved by selecting the auxiliary discharge characteristics in such a way as to satisfy the condition (9). Naturally, one would then have to ensure a certain minimum density of the electron current in the auxiliary gap.

In this case, the possibility of SSVD igniting in a discharge plasma is determined by the condition $T_{dd} = T_0$ (T_{dd} is the auxiliary discharge duration); otherwise, the near-cathode zone can lose electrons before the ionization processes in the discharge plasma are developed.

Figure 4a shows oscillograms of the electron flux current and of the auxiliary discharge current (this was a volume type of discharge) obtained for a mixture of gases of the CO_2:N_2 = 1:3 composition at a pressure of $p = 0.3$ atm when the gap was $d = 2$ cm and the applied bias voltage was $U_0 = 12$ kV. The distance between the auxiliary discharge electrodes was 1 cm and the area of the auxiliary electrode was $S = 200$ cm^2. It is clear from Figure 4a that change in the auxiliary discharge current by an order of magnitude altered $I(t)$ by just 10%, indicating that the condition (9) was satisfied.

The measurements showed that the saturation $I(t)$ is reached at

$$I_{dd} / S = i_{dd} \geq (20-30) i_{\lim} = (20-30) \varepsilon_0 \mu E_0^2 / d, \tag{14}$$

where $i_{\lim}$ is the limiting current density of the primary electron from (12) when Eq. (9) is met. Condition (14) is an empirical criterion, equivalent to (9). Note also that the fulfillment of these conditions becomes simpler when d increases.

Condition (14) can be also easily met when use is made of a pulsed corona discharge as an auxiliary discharge. However, the limiting duration of the stable burning out of both a corona and a volume auxiliary discharge, when condition (14) is met without any special measures taken into account, does not exceed 5–10 μs.

Both volume and corona discharges could be stabilized by utilizing electrodes made of semiconductor materials [18]. Figure 4b shows oscillograms of the volume discharge current in a mixture of gases of the CO_2:N_2:He = 8:32:60 composition with an admixture of triethylamine at p = 1 atm, ignited without a preliminary ionization between a semiconductor material cathode (which was a disk 5 cm in diameter and 2 cm thick) and a grid (discharge gap cathode); it also includes an oscillogram of the charge $q(t)$ of the electrons extracted from the discharge gap in the case when U_0 = 5 kV, d = 8 cm, and the distance between the auxiliary discharge electrodes was 1 cm. The density of the auxiliary discharge current obtained in this way was ~50 mA/cm^2 and the duration of the stable burning of the discharge in gas mixtures of various compositions exceeded 30 μs.

In the case of a discharge of the surface of an insulator, which was limited by the capacitance of the insulator itself, our measurements indicated that the electron current in the gap between the insulator surface and the grid was an order of magnitude less than via an auxiliary electrode. Therefore, (14) is equivalent to the condition

$$| dU / dt | \geq (2-3) \cdot 10^2 \mu E_0^2 \varepsilon_0 / dC_d. \tag{15}$$

Condition (15) is easily fulfilled at $T_{dd} = t_0$ and C_d = 100 pF/cm^2 (C_d is the specific capacity of the dielectric).

Stable burning of long duration discharge, when (14) is met, is realized also in a multichannel discharge grazing along the surface of a semiconductor material limited by the intrinsic resistance of the material due to a choice in the discharge capacitance.

When the necessary condition was satisfied, stable long-term burning of the discharge was achieved also for a multichannel auxiliary discharge on the surface of a semiconductor material, which was limited by the resistance of the material itself, when the charging capacitor was suitably selected. It therefore follows that when special measures were taken, all the types of the auxiliary discharge discussed above were capable of forming an electron flux with the maximum parameters specified by Eq. (12).

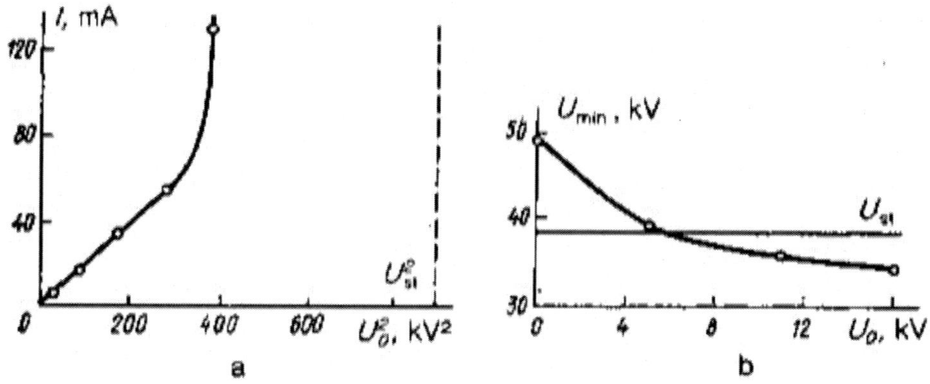

Figure 5. Dependence of amplitudes of the electron flux current (a) and of the minimum ignition voltage of a SSVD (b) on U_0.

Figure 5a shows the dependence of the electron flux current $I(U_0)$, recorded under the same conditions as the oscillograms in Figure 4a. Clearly, at low values of U_0 $I \sim U_0^2$ this dependence is in agreement with Eq. (12). When U0 exceed a certain minimum value I, the current rises steeply and this was accompanied by volume radiation in the discharge gap, indicating the onset of the ionization processes and the formation of a SSVD. This minimum value of U_0 amounted to (2/3) U_{at}, as expected on the basis of Eq. (11), according to which the electric field at the anode was enhanced by the space charge of the electron flux by a factor of 1.5 compared with E_0. We could assume that the possibility of igniting a SSVD at a U_{st} voltage lower than mentioned in Refs. [31] and [21], was also due to distortion of the electric field by the space charge of the electron flux and in Ref. [21] this flux could be formed both from an auxiliary-discharge plasma under the cathode or from the photoplasma in the cathode region, because the discharge on the surface of a ceramic used in Ref. [21] was of relatively short duration (~1.5 μ).

The influence of the distortion of the electric field on the conditions of formation of a SSVD is manifested also by the results in Figure 5b, which gives the dependence of the minimum voltage U_{min} on U_0 for the ignition of such a discharge in the circuit with a preliminary bias of the discharge gap. The sign of this bias was opposite to the sign of U0 in order to avoid formation of an electron flux on application of a voltage pulse to the discharge gap. In this case the auxiliary discharge was of the barrier type. Experiments were carried out on a mixture of the CO_2:N_2:He = 1:1:3 composition and the other parameters were d = 6 cm and p = 0.5 atm. A voltage pulse was applied to the discharge gap after the delay $T = t_0 = d^2/\mu U_0$ relative to the beginning of the auxiliary discharge. It is clear from Figure 5b that U_{min} decreased monotonically on increase in U_0 in accordance with the above model.

At low over-voltages across the discharge gap the distortion of the electric field by the space charge of the electron flux had a considerable influence on the conditions of energy deposition in the plasma of the SSVD and on the residual voltage across the gap after the end of the discharge.

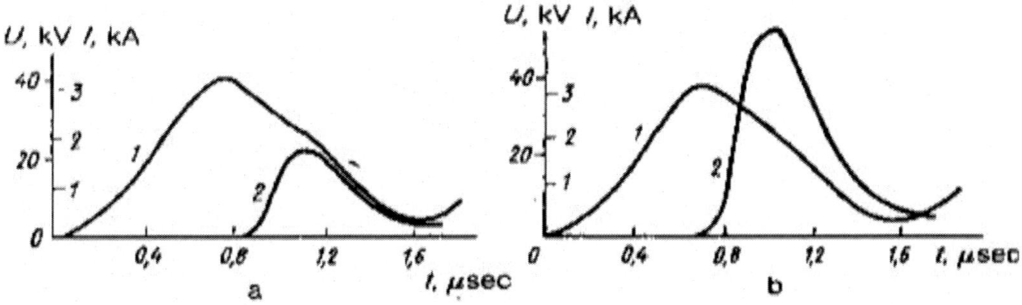

Figure 6. Oscillograms of the voltage U (curves denoted by 1) and of the current (curves denoted by 2) for a SSVD when the condition (9) is not satisfied (a) and when it is satisfied (b).

Figure 6a gives oscillograms of the voltage $U(t)$ and current in a SSVD formed when the density of the electron current was too low for a significant distortion of the field and the condition (9) was not obeyed (d = 6 cm, gas mixture of the CO_2:N_2:He = 1:1:3 composition, p = 0,5 atm, charging voltage U_{ch} = 45 kV, T_f = 0,8 μs). Figure 6b shows oscillograms of $U(t)$ and $I(t)$ recorded under the same conditions as in Figure 6a, but in this case the condition (9) was satisfied. A comparison of Figures 6a and 6b indicated that distortion of the electric field by the space charge of the electron flux resulted in some reduction in the amplitude of the voltage pulse across the discharge gap and also increased the amplitude of the current in the SSVD; it also increased the energy deposited in the discharge by a factor of about 2. This could be important under conditions when it would be difficult to establish a high overvoltage across the discharge gap.

All the forms of the auxiliary discharge investigated by us using a system of profiled electrodes made it possible to form a SSVD when the energy deposited was up to 300 J/liter. The content of CO_2 in CO_2-N_2-He gas mixtures (with the addition of easily ionizable substances) at atmospheric pressure reached 40% when the total content of all the molecular gases was up to 70%, whereas in helium-free mixtures it was up to 30%.

We also established a SSVD in atmospheric air (p = 0,4 atm, d = 10 cm) by the addition of triethylamine, which was evidence of the high efficiency of the method of formation of a SSVD by filling the discharge gap with an electron flux, and this was true also of gases characterized by strong electron trapping.

The best characteristics of a SSVD in a system of non-profiled electrodes (Figure lb) did not differ from the characteristics of the corresponding discharge observed when the initial distribution of the electric field in the discharge gap was homogeneous. The discharge formed using such electrodes could be ignited employing any of the listed types of the auxiliary discharge, with the exception of a volume discharge utilizing metal electrodes (this type of the auxiliary discharge was difficult to realize using non-profiled electrodes because of the inhomogeneity of the electric field in the auxiliary gap). The investigated system was characterized by a strongly inhomogeneous initial distribution of the electric field in the discharge gap, which was close to the field distribution between one of the electrodes and a plane of symmetry in a parallel-plate capacitor of limited size (as shown in cross section in Figure 1b); the field on the anode decreased monotonically along the normal to the symmetry axis of the discharge gap, whereas the field on the cathode increased reaching its maximum at the edge. Consequently, the static breakdown voltage in the system with nonprofiled electrodes was approximately half the corresponding voltage in the quasi-steady phase of a

SSVD. When the discharge gap was subjected to a voltage pulse of amplitude $U > U_{st}$ (representing the regime with a deliberately delayed leading edge of the voltage pulse), a spark breakdown at the cathode edge was observed in the absence of an auxiliary discharge. When the auxiliary discharge $U_{st} < U < 1{,}8U_{st}$ was started and the static voltage was in the required range, a spark did not appear during the period T_{dd} even when the formation of such a spark in the absence of an auxiliary discharge was less than T_f. A weak electron flux current was then observed in the discharge gap. When the amplitude of the voltage pulse exceeded required level and an auxiliary discharge was started simultaneously with the leading edge of the pulse, a volume discharge appeared in the gap.

It therefore follows that the distortion of the initial distribution of the electric field by the space charge of the electron flux in the case of non-profiled electrodes resulted, in contrast to the case of profiled electrodes, in an increase (rather than a reduction) of the pulsed breakdown voltage of the discharge gap and the breakdown was of volume nature. A strongly inhomogeneous initial distribution of the electric field in the presence of non-profiled electrodes was corrected in the process of profiling the discharge gap with electrons, i.e., a dynamic profiling of electrodes by the space charge of the electron flux took place. We shall now qualitatively consider this phenomenon.

Screening of the cathode by the negative charge of electrons reduced several-fold the electric field on the cathode surface, compared with the average field $U(t)/d$ in the discharge gap at a given moment, so that ionization processes could not develop near the sharp edges of the cathode. The field at the anode was then enhanced and the relative magnitude of the enhancement should be highest at the center of the electrode, since the electron flux spread in the course of propagation in a discharge gap as a result of the electrostatic repulsion between the electrons. Bearing in mind also the initial distribution of the field in the discharge gap, we concluded that the ionization processes began to develop simultaneously from the anode to the cathode and from the center to the edges of the discharge gap, i.e., the propagation of an ionization wave, which formed a cathode layer when it closed the discharge gap, was not governed by the initial configuration of the electric field but by that configuration which was formed with the participation of the space charge of the electron flux This made it possible to prevent spark breakdown and to form in a system with non-profiled electrodes a SSVD with a deposited energy distribution typical of such non-profiled electrodes. Measurements of the small-signal gain at low pump energies, which represented the distribution of the energy deposited in the discharge in a plane perpendicular to the optic axis, demonstrated that the gain decreased away from the optic axis to the electrode edges. This provided an indirect confirmation of the above description of the formation of a SSVD. It should be stressed that a change in the voltage on the electrodes resulted in a redistribution of the electric field in the discharge gap at a finite rate governed by the electron mobility. Therefore, the condition for screening of the cathode by the space charge of the electron flux (where U/d is the instantaneous value of the field) could be disturbed if the voltage across the discharge gap rose too rapidly. In fact, it was not possible to create a SSVD using non-profiled electrodes when the voltage was applied abruptly to the electrodes and this was true even when the application of a bias resulted in a preliminary filling of the discharge gap with electrons. Sparks were observed at the corners of the cathode, i.e., in the regions with the maximum concentration of the electric field in the absence of screening. The discharge was stable in the case when the leading front of the voltage pulse was delayed deliberately [21, 22] so that the duration of this front was > 2 μs. for $d = 10$ cm.

A classical example of non-profiled electrodes with a strongly inhomogeneous initial distribution of the electric field in the discharge gap was the first transversely excited CO_2 laser [16]. The present experiments indicated that the electron flux in such a system was formed by a barrier discharge plasma created directly on the cathode surface and this made it possible to satisfy quite readily the saturation (Eq. (9)) and screening conditions. One should stress also that an increase of the stability of a SSVD [22] in systems of the type described in Ref. [16] and considered in the present study, which was observed when easily ionizable substances were added to the gas mixtures, was not due to photoionization of these additives because in the case of regimes used to form a SSVD by filling the gap with an electron flux the contribution of the photoionization to the conductance of the discharge gap during the filling process was unimportant compared with the contribution of the electron flux from the auxiliary discharge plasma and this was true for the majority of the auxiliary discharges. Our investigations made it possible to increase the aperture of SSVD based lasers and amplifiers up to 80 cm and to obtain radiation energy of CO_2 lasers more than 5 kJ for electric efficiency more then 10%.

Chapter 4

HIGH POWER/ENERGY WIDE APERTURE HF/DF LASERS

INTRODUCTION

The main problem arising in the construction of powerful wide-aperture nonchain HF(DF) lasers in which a chemical reaction is initiated by a SSVD is the implementation of the SSVD itself in mixtures of SF6 with hydrogen carriers (deuterium carriers). The possibility of applying for such purposes the usual methods based on the preliminary ionization of the gas in a homogeneous electric field with the subsequent application of a high-voltage pulse to the discharge gap [32] is limited by the high electro-negativity of the SF6 molecules. A new approach to the solution of the problem of increasing the volume of the active medium and the energy of the radiation emitted by nonchain HF(DF) lasers was found in our earlier studies [33, 34, 35]. In a study [34] of a nonchain HF(DF) laser with preionization by soft x-rays, it was observed that the SSVD characteristics and the laser output characteristics are independent of the presence of pre-ionization if the cathode surface has been subjected to a treatment involving the deposition on it of small-scale inhomogeneities. In other words, if the cathode has a rough surface, pre-ionization is not essential in order to obtain an SSVD in the mixtures of an HF(DF) laser.

The possibility of achieving an SSVD without pre-ionization and with an energy input homogeneous throughout the volume under the conditions of a high edge inhomogeneity of the electric field in the discharge gap of an HF(DF) laser has also been demonstrated [35]. These investigations made it possible to increase the aperture of a nonchain HF(DF) laser based on SF6 – hydrocarbon (deuterocarbon) mixtures up to 27 cm and to obtain a radiation energy of ~ 400 J for an electric efficiency of approximately 4% [39].

An SSVD without preionization in SF6 and in mixtures containing it, i.e., a self-initiated volume discharge (SIVD) [36], has a number of properties unusual for volume discharges. The observation of a discharge of this kind opens up unique opportunities for the development of simple and effective nonchain HF(DF) lasers operating in both pulsed and repetitively pulsed regimes. Further investigations of the SIVD in order to establish the principal mechanisms determining its development is therefore of undoubted interest.

To review the results of our investigations is in fact the subject of the present book.

EXPERIMENTS

The SIVD characteristics were investigated in discharge gaps with electrodes having different configurations (Figure 1) with the ratio of the partial pressures of the mixture components SF_6:C_2H_6 = 10:1-10:3 and Pd = 0,02–0,7 cm bar (P is the pressure of the mixture and d the interelectrode distance). The integral characteristics of the discharge and the dynamics of its development were studied using a system of plane electrodes illustrated schematically in Figure 1a. The SIVD was struck with d = 2–6 cm between an aluminum cathode 6 cm in diameter, rounded (1 cm radius) along the perimeter, and an aluminum anode 12 cm in diameter. The surface of the cathode, including the rounded parts, was sandblasted.

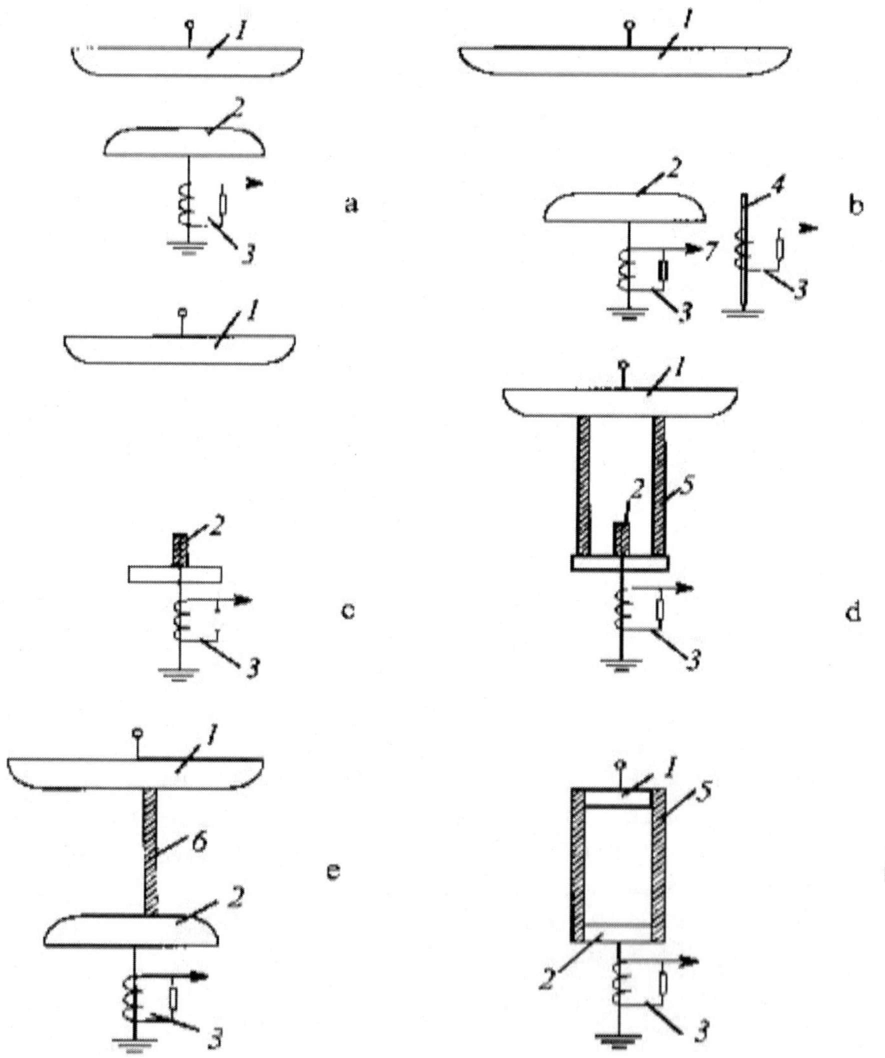

Figure 1. Discharge gaps with electrodes having different configurations: (1) anode; (2) cathode; (3) Rogowski loop; (4) initiating electrode; (5) glass tube; (6) glass plate.

In order to investigate the SIVD dynamics, an artificial (transmission) line with a variable number of cells was discharged across this gap [37]. This made it possible to follow, by varying the duration of exposure to the voltage across the gap in the range 40–400 ns at a constant current, the development of the discharge in time and in space by photographing it for various exposures to the voltage. In the remaining SIVD studies in which the above and other electrode systems shown in Figure 1 were used, the discharge of a capacitor was used to strike the volume discharge. The discharge gaps (Figs 1a and 1b) were additionally illuminated from the side by a low-current ($I < 3$ A) spark enclosed in a quartz envelope. This spark could not ensure the volume photoionization of the gas, but, because of the initiation of electrons at the cathode, it made it possible to stabilize the scatter of the gap-breakdown delay times. The discharge gaps in the configurations shown in Figures 7c-7f were not additionally illuminated.

The dynamics of the development of the SIVD was also investigated in an electrode system with a sectioned cathode (Figure 1b). Here, the main cathode and anode were the same as in Figure 1a, but an earthed conductor 1.5 mm in diameter (the initiating electrode) was connected in parallel to the main cathode at a distance 5 mm in a horizontal plane extending from its edge. The vertical distance between the cathode surface and the tip of the conductor was chosen in such a way that, during the initiation of the discharge by a low-current spark, the electrical breakdown of the gap extended initially to the conductor.

A conductor with polyethylene insulation and a core diameter of 1.5 mm served as the cathode in the electrode system in Figures 1c and 1d (the discharge occurred between the rod and the plane). In these experiments, the SIVD was investigated under conditions such that not more than one cathode spot could be formed on the cathode. In the discharge gap shown in Figure 1d, the discharge cross section on the anode was confined by a glass tube 6-8 mm in diameter (confined discharge), placed between the electrodes in such a way that its upper edge touched the surface of the anode.

In the configuration presented in Figure 1e, the plane electrodes (the same as in Figure 1a) were linked by a dielectric plate 2 mm thick with a transverse dimension of 5 cm. In the electrode system shown in Figure 1f, the SIVD was struck for an interelectrode distance (the distance between two aluminum disks 3.6 cm in diameter) of 5.5 cm; the disks were placed in a quartz tube with an internal diameter of 3.8 cm, the electrodes had sharp edges, and the cathode surface was sandblasted.

The SIVD dynamics was studied in discharge gaps with different geometry. The scheme of the setup for investigating the discharge in a gap with a linear cathode geometry is shown in Figure 1a. The SIVD with a total duration of ~370 ns was ignited in the SF_6: C_2H_6 = 10:1 mixture at a pressure p = 33 Torr and an interelectrode distance of 4 cm. The electrodes were a 0.5-mm thick copper bar 16 cm in length (a knife-edge cathode) placed edgewise and a disk anode 6 cm in diameter rounded over the perimeter with a radius of curvature of 1 cm. The breakdown was forcedly initiated at the center of the cathode with a weak-current spark limited with an R = 900 Q resistor. This spark could not provide an initial electron density in the gas high enough for the development of a volume discharge, but allowed fixing the position of the primary gap breakdown. The SIVD glow was recorded with a single-frame streak camera tube with a frame exposure time of 20 ns, which was triggered with a variable delay T relative to the instant of gap breakdown.

To study the SIVD dynamics in the plane gap geometry, we used the circuit with a sectioned cathode shown in Figure 1b. In this case, the interelectrode distance, the working mixture pressure, and the electrical circuit of the setup were the same as in the previous experiment, except that the cathode was not a copper bar, but a disk 6 cm in diameter rounded over the perimeter with a radius of curvature of 1 cm and subjected to sandblast processing [33, 34, 35, 36]. Insulated conductors 1 mm in diameter were inserted into holes 2 mm in diameter drilled 4 cm apart through the cathode in its plane region. One of the conductors (1) protruded above the cathode surface by 1 mm to ensure the occurrence of the primary gap breakdown at precisely this point. By comparing the oscilloscope traces through the ignitor (7) and monitoring (2) electrodes, we could trace the SIVD spread over the gap.

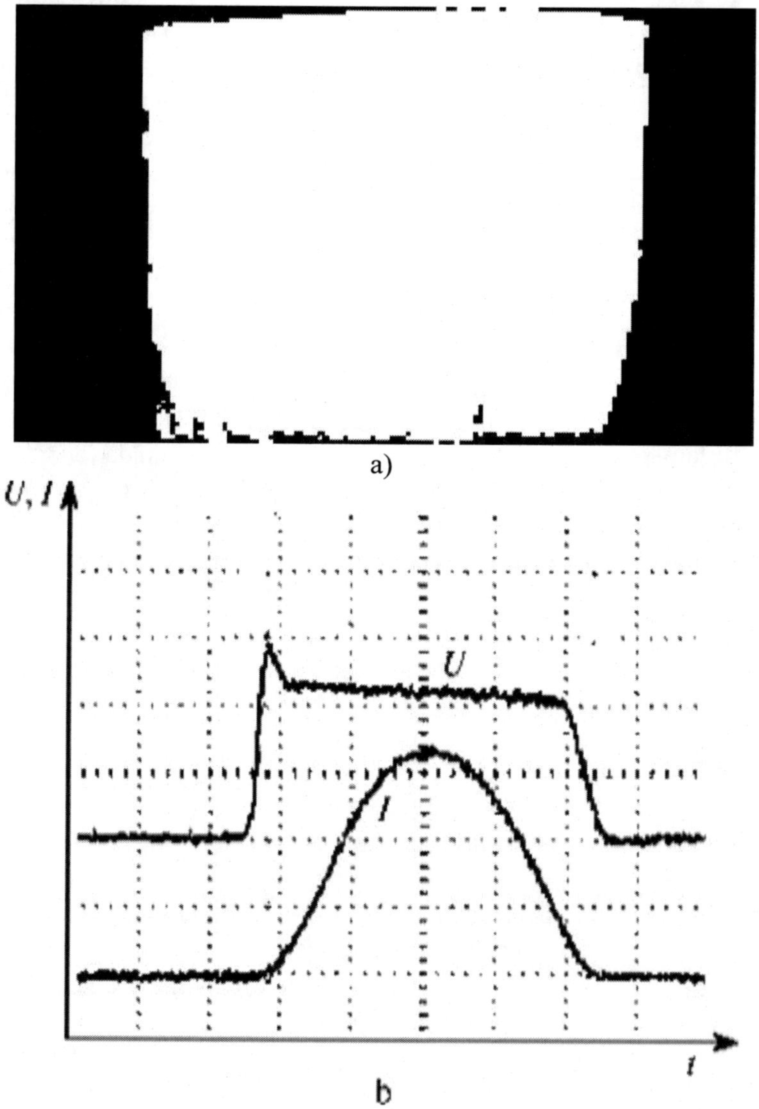

Figure 2. Photograph of an SIVD in a system of plane electrodes (a) and typical current (*I*) and voltage (*U*) oscillograms for the SIVD; scan|100 ns division-[1] (b).

The dynamics of an individual diffusion channel was investigated using the setup shown in Figure 1c. The diffusion channel was simulated by a discharge with a rod (cathode) – plane gap geometry at a pressure p = 16,5 – 49,5 Torr of the SF_6: C_2H_6 = 10:1 mixture for an interelectrode distance of 4 cm. The cathode was the end of a polyethylene insulated wire with a core diameter of 1.5 mm, and the anode was a disk 10 cm in diameter. The SIVD development was monitored with the streak cameras, as in the experiments performed using the setup of Figure 1a.

In the circuits shown in Figs 1a and 1b, the capacitor C discharged into the gap via inductance L upon switching spark gap. In the circuit of Figure 1c, an artificial line discharged into the gap. The currents in the circuits of Figure 1 were monitored with Rogowski coil Rc, the voltage across the gaps was measured with resistive dividers.

A typical photograph of an SIVD in a plane-electrode system is presented in Figure 1a. Figure 1b shows the corresponding current (1) and voltage (U) oscillograms. It is seen from Figure 2 that the SIVD does not outwardly differ from the usual SVD with pre-ionization. It consists of a set of diffuse channels, linked to bright cathode spots and expanding towards the anode. By overlapping, the diffuse channels generate an overall diffuse glow. Despite the appreciable strengthening of the electric field on the edge of the gap, this region is scarcely distinguishable and the cathode spots cover the entire surface of the cathode. The SIVD voltage and current oscillograms are also typical for SSVD in electronegative gases. Energy is deposited in the SIVD plasma at a quasi-steady-state voltage. The voltage in mixtures of SF6 with hydrocarbons, when the content of the latter is not greater than 17%, depends only slightly on their partial pressure. The dependence of U_{qs} on Pd for Pd = 0,05 to 2,5 cm bar and specific energy inputs in the discharge plasma up to 0.2 J cm^{-3} is described satisfactorily by the expression [38] where $U_{qs} = A + B$ $PSF_6\, d$ is the partial pressure of SF6 in the mixture. For pure SF6 and SF6: C2H6 = 10:1, 10:2 mixtures, we have respectively A = 0,72 and 0,79 and 1,1 kV and B = 92,7, 94,8 and 96,4 kV cm bar.

The value of B obtained in the present experiments is close to the known [38] critical reduced electric field strength $(E/p)_{cr}$ = 89 kV/cm bar (the difference is ~8%). The differences between SIVD and SSVD are manifested in the dynamics of their development. Figure 3 presents the results of studies on the dynamics of the development of an SIVD in a plane-electrode system (Figure 7a) for the discharge across the gap of an artificial line with a variable number of cells. As can be seen from Figure 9, in contrast to SSVD, an SIVD is struck initially in the zone of the maximum amplification of the electric field on the gap edge in the form of one or several diffuse channels linked to the cathode spots. The radiation from the discharge in the remainder of the gap is not detected at this instant. The channels formed first initiate the appearance of the subsequent channels and the SIVD spreads over the gap perpendicular to the direction of the electric field at U_{qs} constant voltage, gradually filling the entire gap. The total number of spots on the cathode increases from the instant of the electrical breakdown of the gap virtually in proportion to time, i.e., in proportion to the energy injected into the discharge when account is taken of the experimental conditions.

Since the SIVD develops at a constant current (with the exception of the leading and rear edges) under the experimental conditions described, the current through channels formed earlier should diminish after the appearance of new ones. The characteristic features of the development of an SIVD listed above suggest the operation of current-limiting mechanisms in SF6 and in mixtures based on it, which hinders the transfer of the entire energy through a single channel. It appears that precisely these mechanisms are in fact responsible to a large extent for the existence of the unusual form of the discharge such as the SIVD, including the possibility of achieving it in gaps with a strong edge inhomogeneity of the field.

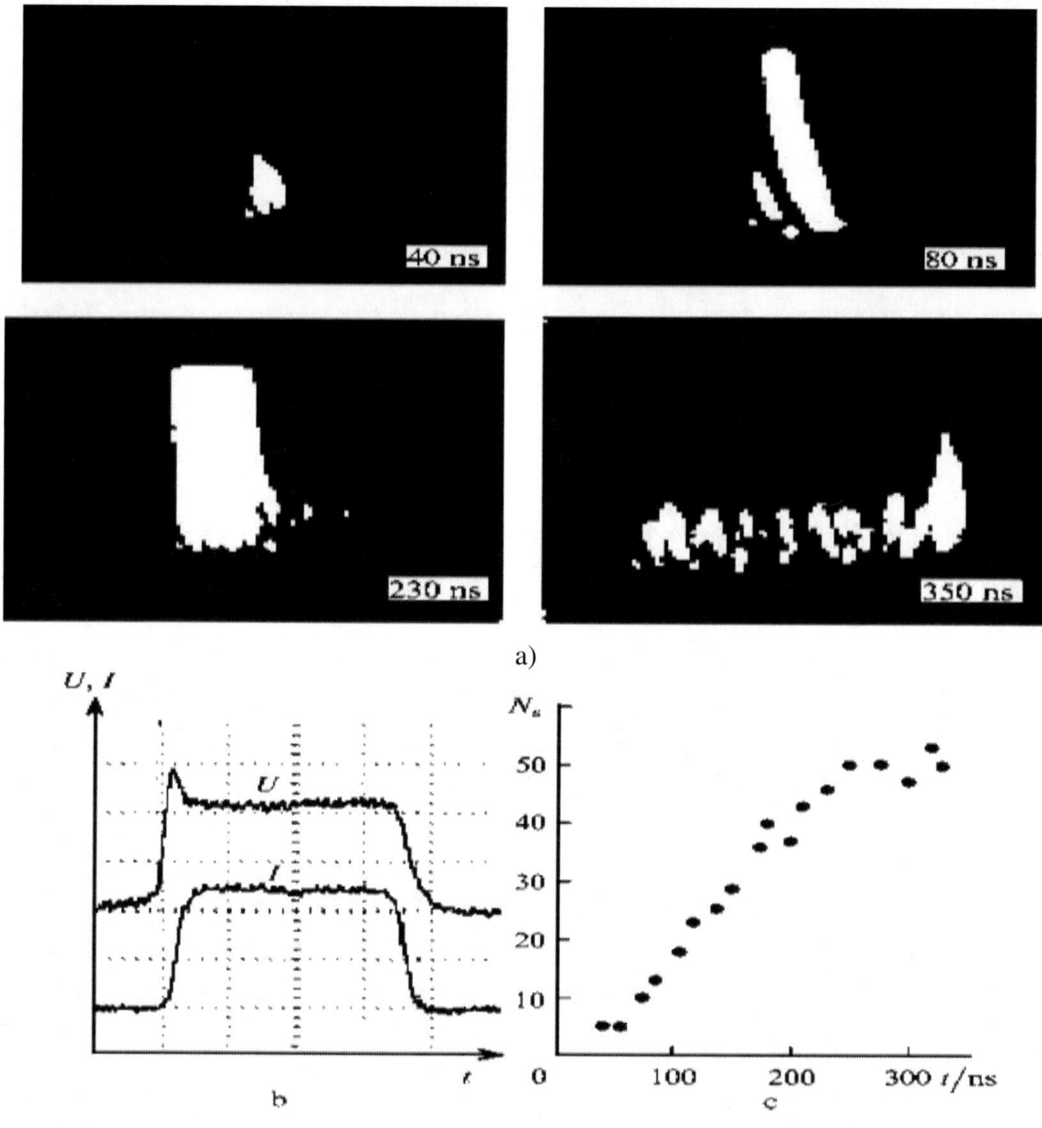

Figure 3. Characteristics of an SIVD: photographs of an SIVD in a system of plane electrodes at different times (spark illumination from the left) (a), voltage (*U*) and current (*I*) oscillograms for the SIVD during the discharge across the gap of an artificial line consisting of 20 cells (scan - 100 ns division [1] (b) and time dependence of the number of spots on the cathode N_s (c).

The existence of such mechanisms is also indicated by the results of experiments in an electrode system with simulation of the first diffuse channel by the initiating electrode (Figure 1b). Figure 4a, which presents characteristic oscillograms of the discharge current through the initiating (curve 1) and main (curve 2) cathodes, shows that the current through the initiating cathode does indeed appear appreciably earlier than the current through the main cathode and its maximum is attained at an instant when the main current still continues to grow.

Figure 4b illustrates the dependence of the fraction of the energy A, transmitted by the initiating cathode, relative to the total energy W injected into the discharge. The quantity A decreases monotonically with increase in W, reaching saturation, which yet again constitutes evidence in favor of the existence of mechanisms preventing the transfer of the entire SIVD energy through one channel.

However, we may note that that in the experiment under consideration, the simulation of the initial diffuse channel with the aid of a special electrode is not entirely equivalent to the real conditions, because, in order to guarantee that the discharge begins from the initiating cathode, it would be necessary to raise it by approximately 1 mm above the surface of the main cathode. For this reason, the brightness and transverse dimensions of the simulated initial diffuse channel differed appreciably from those of channels in the main gap, whereas under real conditions this was not observed (Figure 2).

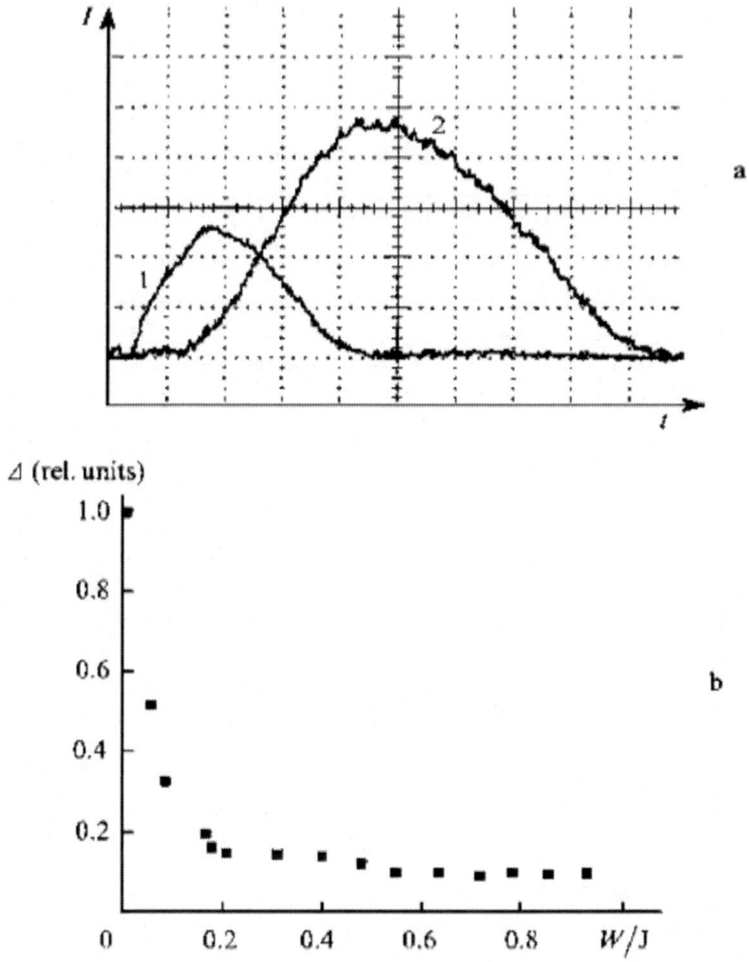

Figure 4. Oscillograms of the current through the initiating (1) and main (2) cathodes (39.7 A division, scan 50 ns division)- (a) and dependence of the fraction of energy A transmitted through the initiating electrode relative to the total energy W (b).

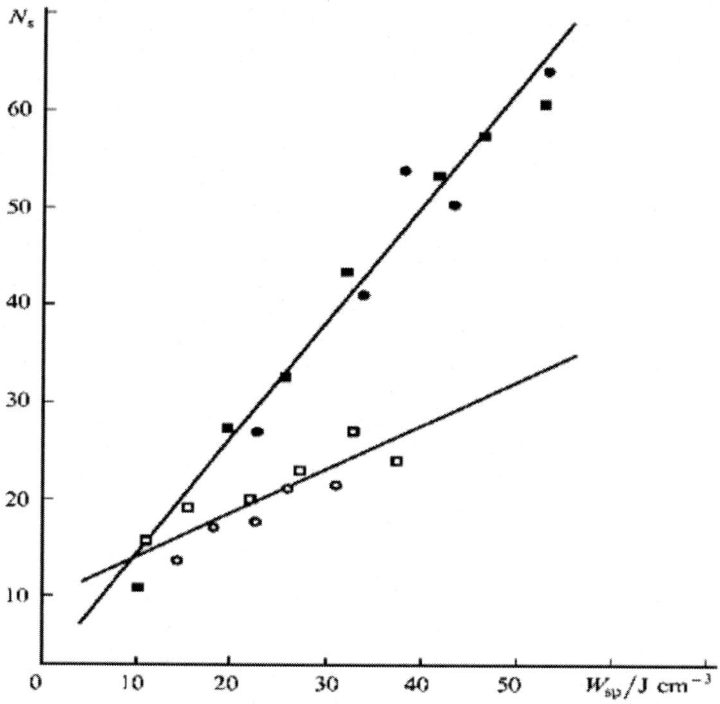

Figure 5. Dependence of the number of cathode spots N_s on the specific input energy W_{sp} for the SF6: C2H6 = 10:1 mixture with: d = 6 cm and P = 33.6 Torr (.); d = 2 cm and P = 33.6 Torr (); d = 6 cm and P = 16.8 Torr (n); d = 2 cm and P = 16.8 Torr (o).

Figure 5 presents a typical dependence of the total number of spots on the cathode N_s on the average specific energy W_{sp} injected into the plasma. It was recorded in a system of plane electrodes for different values of d and p. W_{sp} was varied during the experiments by varying the charging voltage and charging capacitance. The duration of the discharge current was also varied. Evidently, for a fixed value of P, N_s depends only slightly on d. Within the limits of error, the points satisfactorily on a linear dependence of N_s on W_{sp}.

Consequently, for a fixed value of p, the number of spots on the cathode is determined by W_{sp} and not the density and duration of the current, W_{sp} actually specifying the total number of spots on the cathode. For a short duration (less than 100-150 ns) of the current, the spots may not cover the entire surface of the cathode but only its section near the site of the primary electrical breakdown of the gap. In short discharges, the local density of spots may therefore be greater than in long discharges.

The dependence of N_s on p may be more complex: N_s increases with increase in P and the increase is nonlinear. A diminution of the transverse dimensions of the spots on the cathode and of the diffuse channels on the anode is also observed under these conditions, i.e., the volume occupied by each channel decreases simultaneously with increase in N_s.

One may postulate that such influence of the growth of p on N_s is associated with the increase in the electric field strength (defined as $Eqs = Uqs/d$) for which an SIVD develops and which is known [1] to determine the probability of the formation of a cathode spot.

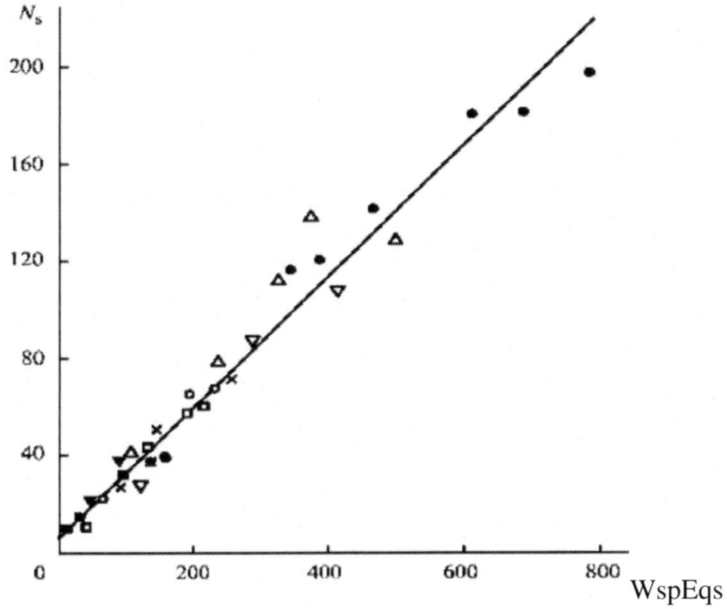

Figure 6. Dependence of the number of cathode spots *Ns* on *WspEqs* for the SF6: C2H6 = 10:1 mixture with: *d* = 6 cm and *P* = 33.6 Torr (n); *d* = 6 cm and *P* = 16.8 Torr (); *d* = 4 cm and *P*= 23.3 Torr (*); *d* = 3 cm and *P* = 23.3 Torr (); *d* = 3 cm and *P* = 33.6 Torr (o); *d* = 3 cm and *P* = 50.4 Torr; *d* = 3 cm and *P* = 67.2 Torr (); *d* = 2 cm and *P* = 33.6 Torr (x).

The dependence of N_s on the parameter *Wsp Eqs* plotted for different values of *P* and *d* taking into account this hypothesis, is presented in Figure 6. It is satisfactorily fitted by the linear function $N_s = A + B\ WspEqs$, in which the constant *b* is in its turn a function of the state of the cathode surface and of the hydrocarbon content in the mixture [39].

A conclusion important for the further interpretation of the mechanisms of the development of an SIVD follows from Figures 5 and 6: the volume occupied by the discharge increases with increase in the amount of energy injected into the plasma and almost in proportion to the energy.

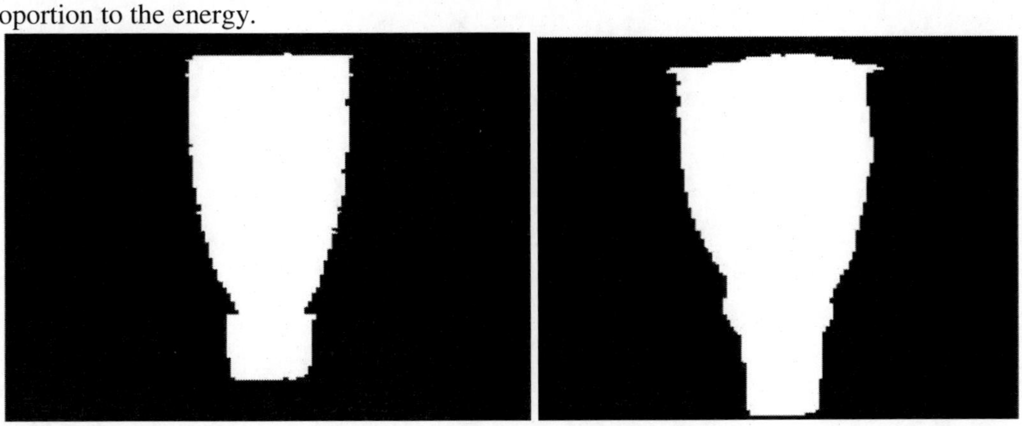

a) b)

Figure 7. Photographs of an SIVD in the rod-plane electrode system with $W = 0.73$ J (a) and 2.33 J (b).

Figures 7a and 7b present photographs of an SIVD in a rod-plane electrode system (Figure 7c) for d = 5 cm and P = 34 Torr in the SF6: C2H6 = 10:1 mixture recorded for two energies W deposited in the discharge plasma. In this geometry, the SIVD consists of a diffuse plume, expanding towards the anode, with formation of bright plasma on the cathode. Under conditions such that new spots cannot be formed on the cathode, the volume V of the glowing part of the plume (an analogue of a diffuse channel) increases with increase in W.

The dependence of V on W obtained under the same conditions as in Figure 7 is presented in Figure 8. A linear growth of V with increase in W can be seen. For W = 5 J, the volume V is ~ 20 cm^3, whereas the size of the plume on the anode is 3.6 cm. The high stability of the discharge in the gap geometry under consideration, despite the fact that the current density of the cathode is j = 1.2 10^5 A cm^{-2}, is striking. It is noteworthy that the influence of the duration of the discharge current on V is also appreciable in this geometry. This accounts for the considerable scatter of the points in Figure 8 because in the experiments the duration of the current varied from 50 to 300 ns.

Figures 9a and 9b present voltage (U) and current (I) oscillograms for an SIVD in the rod-plane electrode system (Figure 1c) for current pulses of 100 and 200 ns duration with identical charging voltage and charging capacitance in the SF6: C2H6 = 10:3 mixture (P = 40 Torr, d = 4.4 cm). Figs 15c and 15d show similar voltage and current oscillograms for an SIVD recorded under the same conditions but with confinement of the discharge on the anode by a tube 7 mm in diameter (Figure 1d). As can be seen from Figs 9a and 9b, the voltage and current oscillograms in the rod - plane system do not differ in any way from typical oscillograms obtained in a plane electrode system.

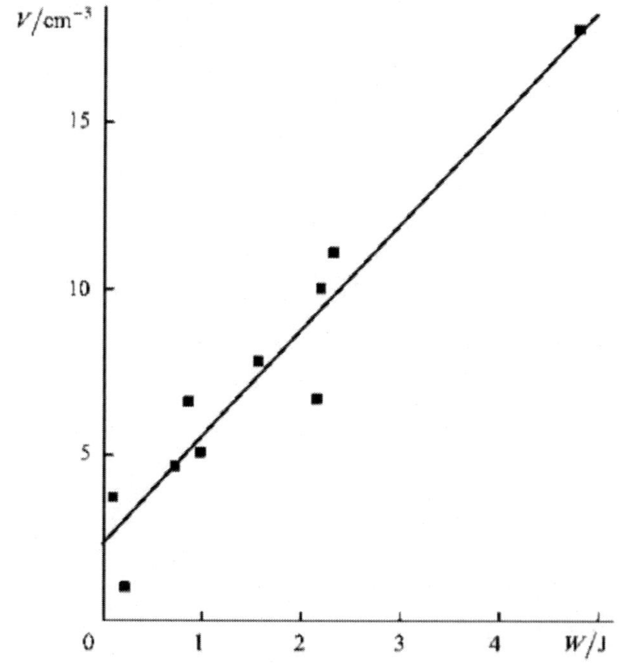

Figure 8. Dependence of the volume V of an SIVD on the energy W in the rod - plane electrode system.

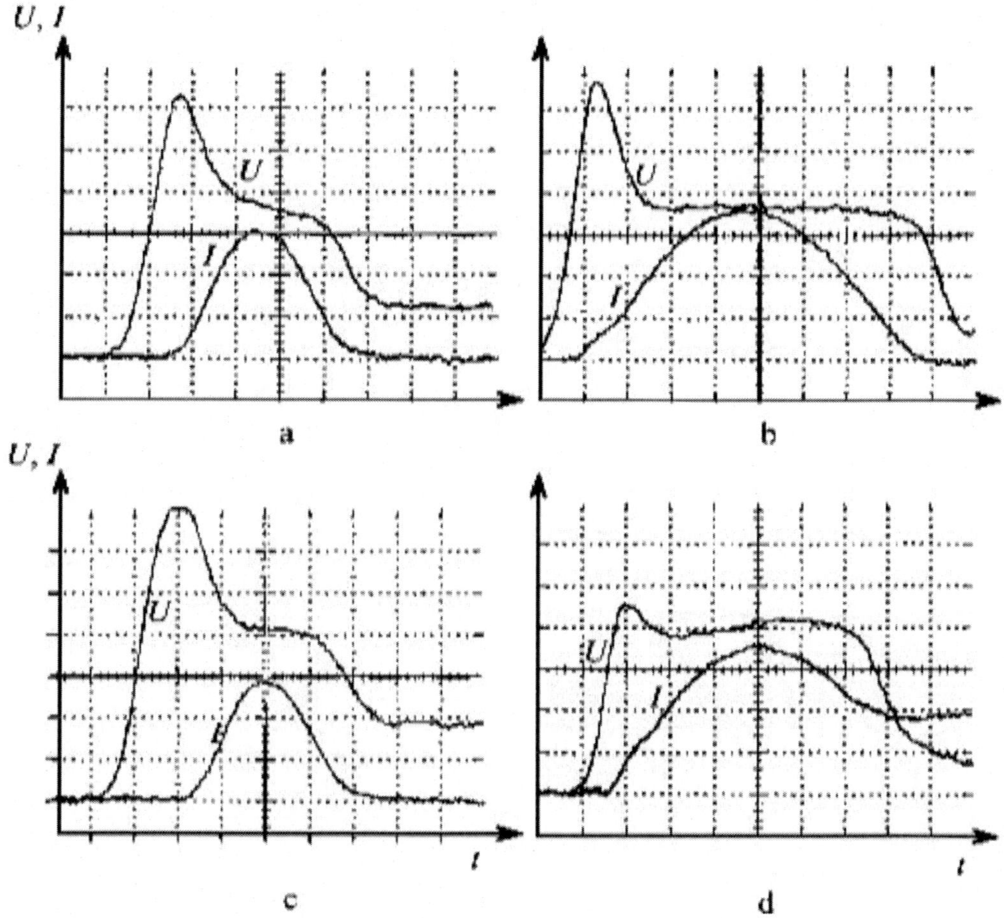

Figure 9. Voltage (U) and current (I) oscillograms for an unconfined (a, b) and confined (c, d) SIVD in the rod-plane electrode system with T = 100 ns (a, c) and 200 ns (b, d). Oscillogram scales: 5.1 kV division (a-d), 79.4 A division (b, d), 198.5 A division (a, c); time scan - 25 ns division.

Confinement of the discharge leads to an appreciable change in the type of oscillograms. For a long duration of the current pulse (Figure 9d), after the initial fall of the voltage, one observes its increase simultaneously with increase in the current, the current maximum being attained sooner than in the case of an unconfined discharge. The current amplitude also diminishes, whereas the voltage is appreciably higher at its maximum than in the case of an unconfined discharge. Voltage oscillograms of a similar type are obtained when the discharge characteristics are calculated taking into account the attachment of electrons to vibrationally excited SF6 molecules [40]. However, the calculated oscillograms in Ref. [40] do not agree with the experimental ones. We are aware of other studies in which a similar form of the discharge voltage was observed in SF6 and in mixtures based on it.

For T = 200 ns, the confined SIVD contracts after the current has reached a maximum. A decrease in the duration of the current makes it possible to obtain an uncontracted confined SIVD (Figure 9c). In this case, there is no break in the voltage oscillogram and, following a general increase in the discharge voltage, the 'tail' becomes less tilted. We may note that the modification of the voltage pulse observed in Figures 9c and 9d following the confinement of the SIVD is associated precisely with the increase in the specific energy input in the plasma because under the conditions of V.V. Apollonov/Central European Journal of Physics 4,708, (2003) the discharge cannot expand during the energy deposition process.

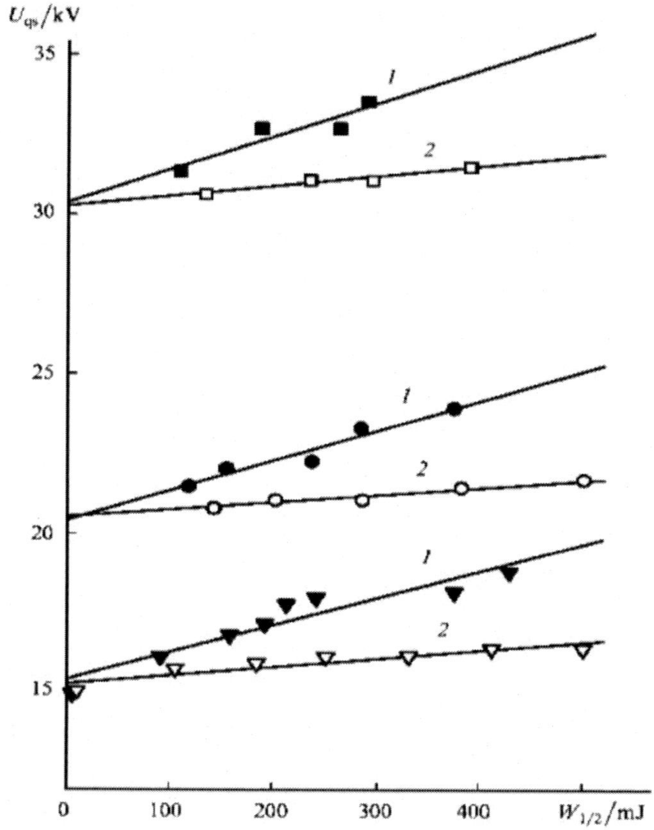

Figure 10. Dependence of U_{qs} on the energy for confined (1) and unconfined (2) SIVD with $P = 33.6$ Torr (triangles), 45.8 Torr (circles) and 67.2 Torr (squares); SF6: C2H6 = 10:1 mixture, $d = 4$ cm.

Figure 10 presents dependences of U_{qs} on the energy Wj/2 deposited in the discharge plasma before the attainment of the current maximum, in the rod-plane electrode system for confined (curve 1) and unconfined (curve 2) SIVD; was determined at the current maximum. Evidently, with increase in *Wj*/2 an appreciable growth U_{qs} is observed in the confined SIVD, significantly more notable than in the unconfined SIVD. The relative increase U_{qs} (compared with U_{qs} for the unconfined SIVD) for the same values of *Wj*/2 greater the smaller the value of *P*, i.e., the greater the energy deposited in the plasma per unit volume and per molecule.

The results presented in Figs 9 and 10 indicate an increase in the electron attachment coefficient in SF6-based mixtures with increase in the specific energy input in the discharge plasma.

The possibility of obtaining a volume discharge in the electrode systems illustrated in Figures 1e and 1f is a consequence of the existence of mechanisms restricting the SIVD energy transmitted by an individual diffuse channel and leading to an increase in the volume occupied by the discharge with increase in the input energy. Figure 11 presents photographs of a discharge in air and in the SF6:C2H6 = 10:1 mixture, respectively, at a pressure of 65 Torr obtained in the system comprising plane electrodes linked by a dielectric plate (Figure

1e). As was to be expected, the discharge in air develops in the form of the electrical spark breakdown of the gap along the surface of the dielectric plate, the bulk phase being entirely absent. In a SF6 - C2H6 mixture, the discharge is very much of the volume type, its external appearance and oscillograms being completely indistinguishable from an SIVD obtained in the same system in the absence of a plate (with the exception that the photograph shows a black band at the site where the plate is located).

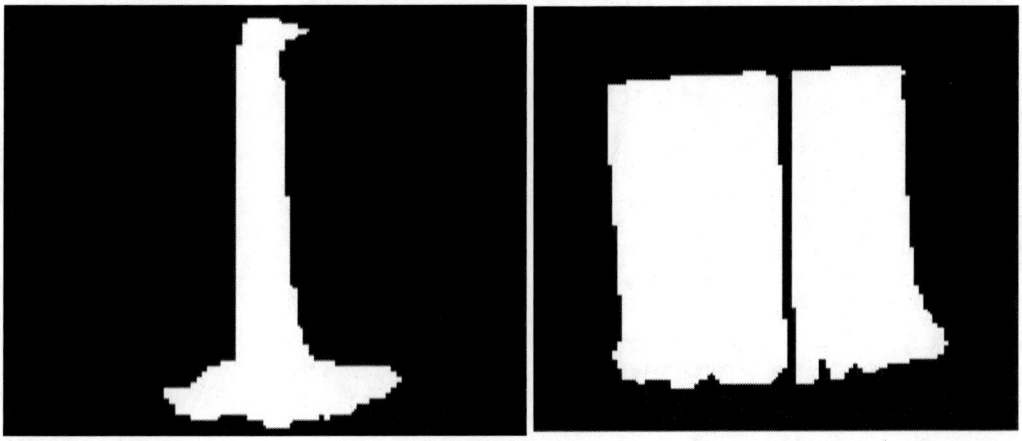

(a) (b)

Figure 11. Photographs of a discharge in a system of plane electrodes linked by a glass plate in air and in the SF6: C2H6 = 10:1 mixture (b).

In the system with transverse dimensions of the electrodes close to the internal diameter of the tube (Figure 1f), the discharge is also very much of the volume type for the mixture SF6: C2H6 = 10:1 and P = 15-65 Torr. When $W_{sp} > 0.05$ J cm^{-3}, the surface density of the cathode spots has a maximum in the central zone of the cathode, as in the usual gap with plane electrodes which are not linked by a dielectric surface.

In the subsequent ignitions of an SIVD without change of mixture, an appreciable increase and a decrease in the current are observed in the above electrode system. Since in the case under consideration the SIVD occupies the entire volume of the discharge chamber (Figure 1f), this fact indicates that, in the mixture of gases formed as a result of the dissociation of the starting materials, the effective ionization coefficient is smaller than in the initial mixture. We may note that a volume discharge in a tube with an internal diameter close to the diameter of the electrodes had been obtained earlier in a SF6 - H2 mixture [41], but the authors [41] did not interpret this result.

DISCUSSION

Thus, the following processes are characteristic of SIVD: consecutive formation of diffuse channels (with a simultaneous decrease in the current through the channels which had arisen earlier) during the development of the SIVD in a gap with planar geometry, the expansion of the diffuse-glow zone in the rod - plane gap, an increase in the discharge voltage with increase in the specific energy. As already mentioned, these processes are apparently determined to a large extent by the mechanisms of the confinement of the current in the conducting channel, which depends on the specific energy input. We shall consider two possible mechanisms leading to this state of affairs, namely the dissociation of SF6 by electron impact and the attachment of electrons to vibrationally excited SF6 molecules.

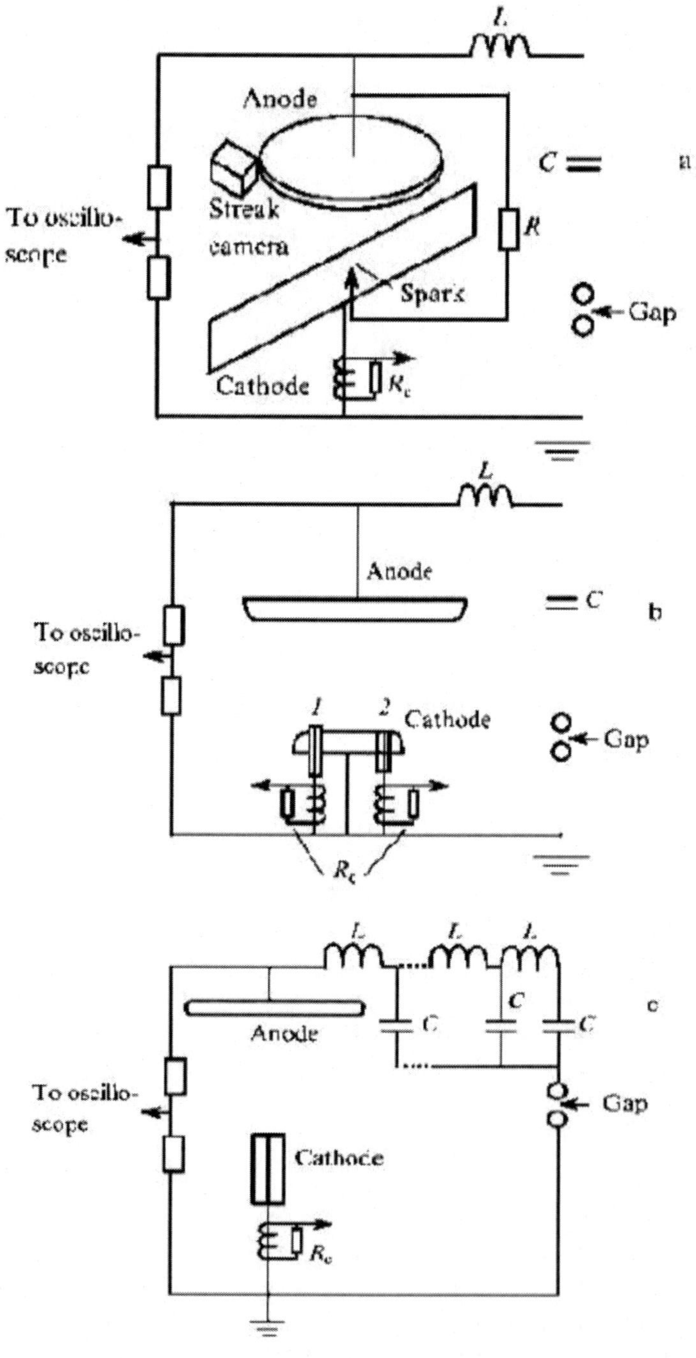

Figure 12. Schemes of experimental setups for studying the SIVD dynamics: ignitor (1) and monitoring (2) conductors.

In the first place we note that, for the ratios *E/N* close to the critical value, virtually the entire energy deposited in SF6 is consumed in dissociation [45]. According to Ref. [38, 42], the SF6 = SF4 + 2F channel is dominant.

The ionization potential of *F* atoms (17.42 eV) exceeds that for SF6 (15.7 eV) [43], so that an appreciable contribution by the *F* atoms produced to the overall ionization should not be expected even at their considerable concentrations. As regards the attachment of electrons, the formation of F^{-1} ions via any of the possible mechanisms is known to be incapable of competing, as shown by estimates [44], with the attachment of electrons to SF6 molecules. The F excitation threshold is 12.7 eV [43] (the component of the main doublet with a threshold energy of ~0.05 eV may naturally be disregarded in the given instance), i.e., it is in the region of the intense excitation of the SF6 terms by electron impact [45]. Furthermore, there are grounds for the postulate that inelastic processes with participation of SF4 molecules also do not influence in any way significantly the energy spectrum of the electrons in the discharge.

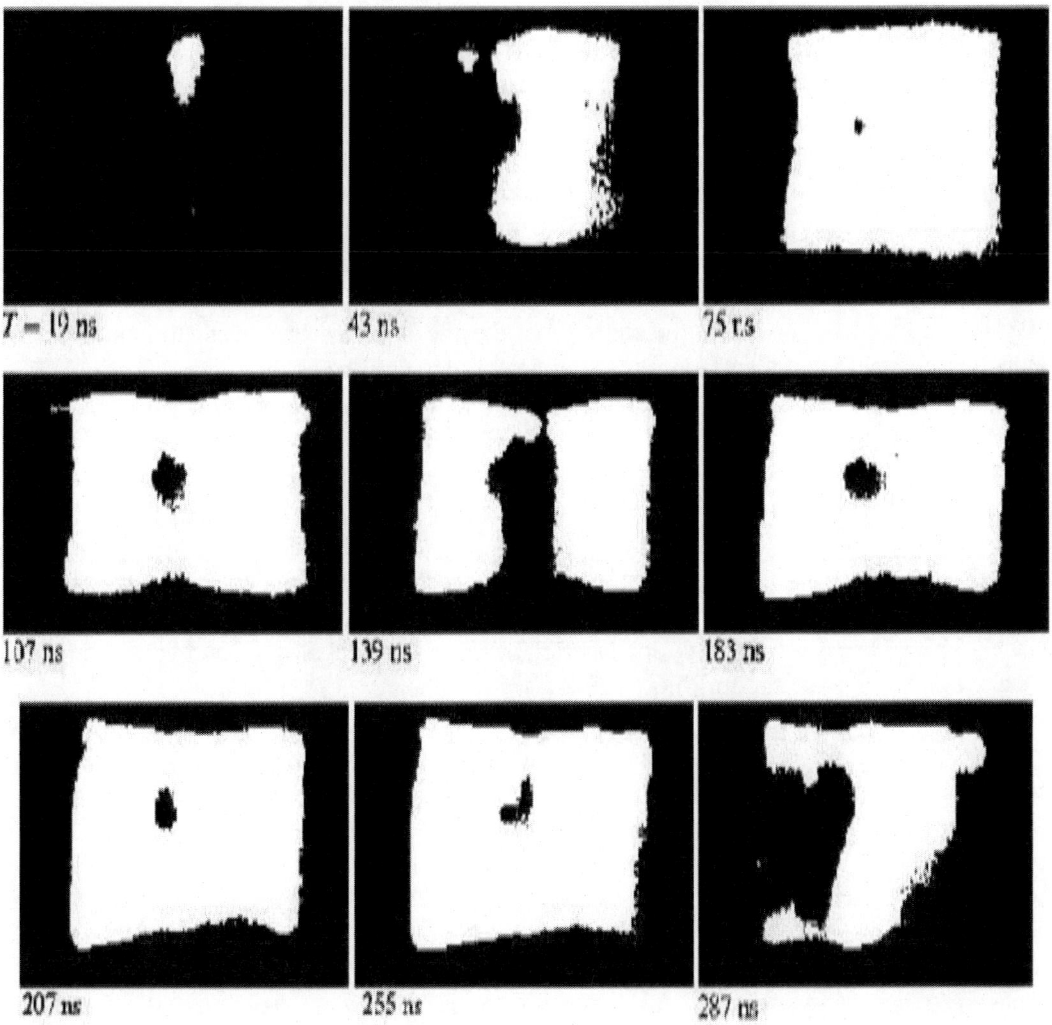

Figure 13. Streak camera images obtained at different instants of time.

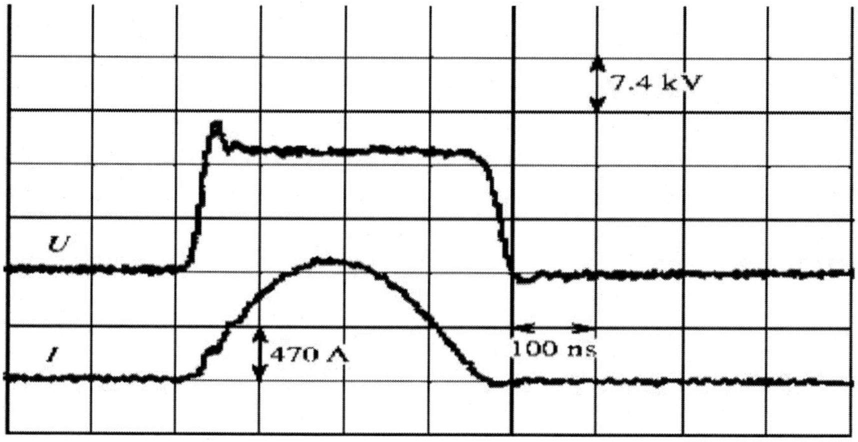

Figure 14. Oscilloscope traces of the voltage and current of SIVD.

Thus, bearing in mind the qualitative nature of the models considered, and neglecting therefore the difference between the cross sections for the elastic scattering of electrons by SF6 molecules and their decomposition products, one may postulate that the influence of the dissociation of SF6 may only affect the decrease in the reduced field strength *E/N* with increase in the total concentration of neutral particles *N* as the specific energy input increases.

It is possible to show that allowance for additional factors will significantly complicate the description without altering fundamentally the result: the dissociation of SF6 by electron impact and the attachment of electrons to vibrationally excited SF6 molecules may indeed serve as the mechanisms of the limitation of the current in the conducting channel in the active media of an HF laser. At the same time, the mechanisms of current limitation in the conducting channel are not sufficient for complete understanding of the processes observed in the experiment, in particular the propagation of the discharge into the interior of the gap in the direction perpendicular to the applied field [18]. Figure 13 shows the streak-camera images of a SIVD in the gap with a knife-edge cathode obtained at different instants of time. The corresponding oscilloscope traces of discharge current and voltage are given in Figure 14. One can see from Figure 13 that the gap is broken down at the center (at the region of the auxiliary spark discharge). At this instant of time, the SIVD has the form of one diffusion channel, with a cathode spot already formed. Then, new channels appear near the first channel, which have a significantly lower brightness, and are developing at a voltage close to the static breakdown voltage for SF6.

The number of new channels increases with time, they 'disperse' from the middle to the periphery of the gap, and their brightness gradually achieves that of the first channel, the channels closest to the first channel being the brightest. All the channels become equally bright with time, while the brightness of the first channel decreases significantly. Upon a further SIVD development, the brightness of the channels at the gap periphery (i.e., the channels removed from the primary breakdown region) gradually increases. However, for $T >$ 210 ns the glow becomes uniform again over the cathode length, the glow in the first channel also being restored. This effect will be referred to as a current return (to the channel). Then, we observed the development of discharge instability against the background of the total diffusion glow.

Therefore, the results outlined above confirm directly the fact that the SIVD spreads over the gap at a voltage close to the static breakdown voltage in the form of a sequence of diffusion channels. In this case, the first channel, which was produced upon the gap breakdown, fades away as new channels emerge. Note also that the current return (see Figure 13), which is manifested in the equalization of the brightness of all the channels and in the increase in the brightness of the first channel following its initial decrease, is observed only when the energy introduced into the plasma is high enough. This effect vanishes when the energy is lowered or the linear size of the cathode is increased. For high energy inputs (150-200 J/L) typical of HF (DF) lasers, the SIVD spreading over the gap is so fast that the time resolution of our streak camera is insufficient A similar pattern of SIVD development is also observed with the plane discharge gap geometry.

Figure 15 shows the oscilloscope traces from the igniter and monitoring conductors (see Figure 12b) obtained in the experiment with the sectioned cathode. One can see from Figure 15 that the current begins to flow through the monitoring conductor with a significant delay relative to the current through the ignitor.

By the time of onset of the current through the monitoring conductor, the amplitude of the current through the ignitor lowers by more than a factor of two, i.e., in this experiment, too, we observed quenching of the initially produced channel upon the emergence of subsequent channels. One can also see from the oscilloscope trace of the current through the ignitor (Figure 15) that a current returns to the first channel after its almost complete quenching in the case of a plane gap as well, provided the energy input is high enough. In the rod-plane gap, the spreading of the frame, which simulated a single diffusion channel, was observed to occur in two stages. During the first 20-30 ns following the gap breakdown, the discharge volume rapidly increased, out of proportion to the energy inputted. Then, the discharge volume increased approximately linearly with the energy.

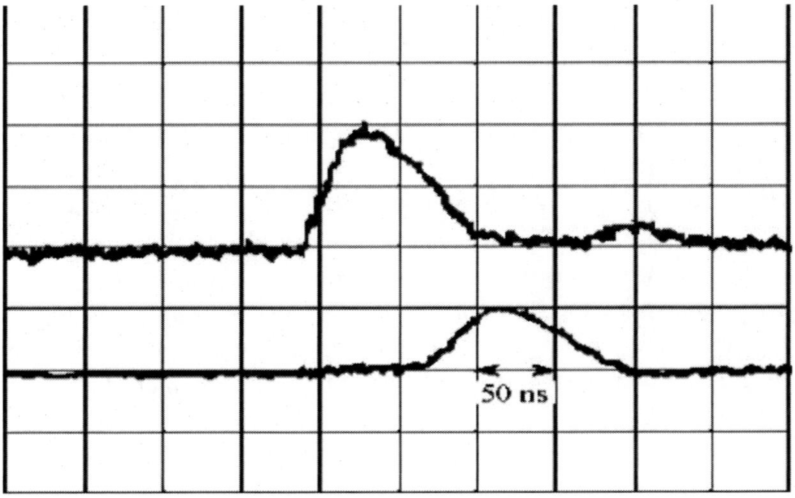

Figure 15. Oscilloscope traces of the current through the ignitor (the upper trace) and monitoring (the lower trace) conductors.

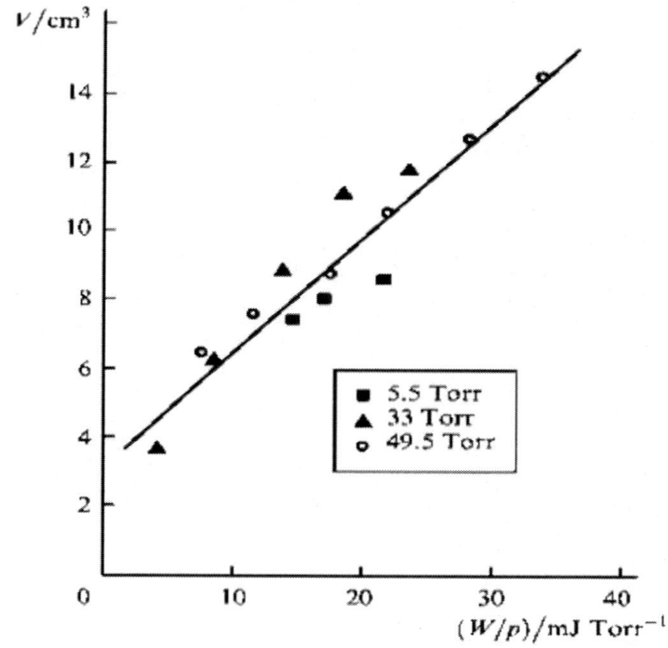

Figure 16. Dependences of the volume *V* occupied by discharge on the parameter *W/p* for the SF6: C2H6 =10:1 mixture at a pressure *P*.

For a constant total energy *W* introduced into the plasma, the largest volume *V* occupied by the SIVD by the instant of energy input termination increased with lowering the mixture pressure *p*. Figure 16 shows the dependence of *V* on the parameter W/P obtained for different *P*. One can see that this dependence is satisfactorily within the error of experimental determination of *V*, described by a linear function.

The results presented above confirm assumptions that there exist mechanisms of current density limitation in the diffusion channel of a discharge in SF6 and SF6-based mixtures [32, 36, 46, 47, 48, 49] (and also in several other strongly electronegative gases [36, 49]). It would be reasonable to attempt to relate these mechanisms to the specific feature of SF6, namely, with its strong electronegativity.

Strongly electronegative gases have highly reduced electric field strengths of *E/N*, where *E* is the field strength and *N* is the concentration of neutral molecules, at which an electrical breakdown occurs and the energy is inputted into the plasma. They also have large electronegativity parameters Xa defined as the ratio of concentrations of negative ions and electrons. For this reason, the following features of the charge kinetics inherent in these gases should be taken into account in the study of a volume self-sustained discharge in strongly electronegative gases such as SF6 and SF6-based laser mixtures.

First, the electron-impact dissociation of SF6 molecules and other components of the mixture plays a significant role in this discharge. Indeed, for high E/N close to the critical value $(E/N)_{Cr}$ = 360 Td, the average energy of plasma electrons in the quasi-stationary stage of a self-sustained discharge approaches the SF6 dissociation threshold and more than 80% of the energy inputted into the discharge is spent to the dissociation [44, 52] resulting in the production of an *F* atom. In the sub-microsecond range typical for the discharge under study, decomposition products have no time to escape from the discharge channel, resulting in a local increase in the concentration N of particles, a decrease in E/N, and a decrease in the electric conductivity of the diffusion channel due to the increase in electron losses through attachment.

High negative ion concentrations are attained in the discharge, with the effect that the electron detachment from negative ions can make a significant contribution to the balance of charged particles. Usually the disintegration of negative ions in their collisions with neutral molecules and/or due to associative ionization is taken into account. However, for medium gas pressures (p = 10 to 100 Torr) and a sub-microsecond discharge duration, these processes do not make any noticeable contribution to electron multiplication [53]. However, there are firm grounds to believe that the electron-impact detachment of electrons from negative ions can be an efficient channel of the delivery of secondary electrons to the discharge plasma [54]. Third, a significant influence on the parameters of an SF6-discharge plasma, including its conductivity, can be exerted by dissociative electron-ion recombination because high concentrations of positive ions are achieved in the discharge, and the energy is mainly inputted into the plasma. Ion-ion recombination must also be taken into account, because it may significantly limit the concentration of ions in the plasma.

Therefore, it follows from the above discussion that the main mechanisms of SIVD current density limitation are the dissociation of SF6 molecules by electron impact resulting in a decrease in the ionization rate and an increase in the electron attachment rate due to the local reduction of the parameter E/N in the diffusion channel, and also electron -ion recombination, which is responsible for the growth of electron losses with increasing current density in the channel. It was shown qualitatively [49] that the current density can be also limited due to the attachment of electrons to vibrationally excited SF6 molecules. However, the absence of reliable data on the rate constants for these reactions in the literature complicates a quantitative estimate of the role of this process in the balance of charged particles in the plasma.

We performed SIVD simulations to verify whether the above assumptions are consistent with the available experimental data. The channel structure of the discharge was modelled with a set of resistive elements connected in parallel. Similarly to Ref. [55], the concentrations of particles were determined by solving continuity equations for particles of each sort in combination with the Kirchhoff equations for the discharge circuit. The nonuniformity of initial conditions for the development of channels along the cathode length was simulated by prescribing different initial electron concentrations in each of the channels. In addition to the electron-impact ionization of SF6 and electron attachment to SF6 molecules, the following processes were included in the calculation.

1) SF6 dissociation by electron impact. The number of dissociated molecules was defined as $Nd = W/qF$, where W is the energy inputted into the discharge and qF = 4.5 eV is the energy spent to produce an F atom [41].
2) Electron detachment from negative ions by electron impact. It was assumed that SF6 - negative ions dominate in the plasma because the charge exchange reactions have no time to occur during the discharge period, while the cross sections for the electron-impact production of other negative ions are too small [52, 53]. The rate constant for the disruption of negative ions by electron impact should not be smaller than the rate constant for elastic electron scattering by molecules [54].

3) Dissociative electron - ion recombination. The rate constant for this process was estimated assuming that the SF6 discharge plasma is dominated by SF_5^+ ions (52).
4) Ion - ion recombination. In this case, the rate constant for the *E/N* values close to the critical value was taken from Ref. [53].

a)

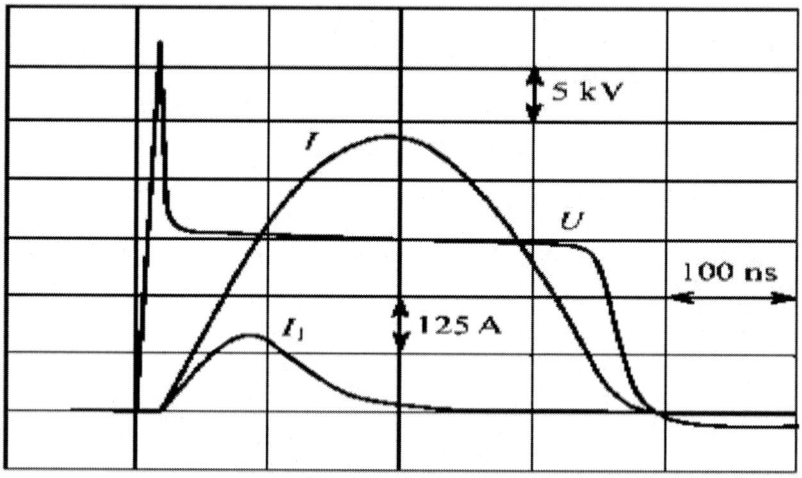

b)

Figure 17. Calculated oscilloscope traces of the voltage *U*, the total discharge current *I*, and the current In through the first channel for energy density 80 (a) and 40 J/L (b).

We simulated an SIVD with nine channels, which corresponds approximately to the conditions of the experiment performed using the setup of Figure 12a. Figure 17 shows the oscilloscope traces of the voltage, the total discharge current and the current through a single channel for which we prescribed the maximum initial electron concentration, calculated for the energy density inputted into the discharge plasma equal to 80 and 40 J/L. One can see from Figure 17a that the current through a single channel has two maxima when the energy input is high. This is in qualitative agreement with the oscilloscope trace of current through the ignitor (see Figure 15) in the plane gap geometry and in agreement with the experimentally observed redistribution of the channel glow intensity upon the SIVD development in the knife edge-plane gap (see Figure 13). The current return vanishes when the energy input is lowered (Figure 17b).

Calculations also show that the electron detachment by electron impact and electronion recombination virtually compensate each other within the accuracy of estimates of the rate constants. The dissociation of SF6 molecules and other components of the HF(DF)-laser mixture should therefore be considered as the principal mechanism responsible for current density limitation in the SIVD channel. However, it is reasonable to refine the magnitudes of the rate constants. Note that our model (like any zero-dimensional model, which neglects the time variation of the SIVD volume) pretends only a qualitative illustration of the current redistribution in the channels when account is taken of possible mechanisms of current density limitation.

Note also that the electron detachment from negative ions by electron impact should be taken into account in the analysis of the processes determining the instability development of the volume discharge in SF6 and other strongly electronegative gases.

Once again, we draw attention to the result that follows from the investigation of the dynamics of a single diffusion channel: an SIVD develops at an approximately constant value of the energy parameter *W/(Vp)*.

High-Power Self-Controlled-Volume-Discharge Based Molecular Lasers

We have investigated in the past the formation of a self-sustained-volume-discharge (SSVD) for CO2–N2–He mixtures by preliminary filling the gap with an electron flux from an auxiliary discharge plasma source and self-initiated-volume-discharge (SIVD), which is the most interesting due to wavelength for the topic under consideration. The first method is suitable for the large distances between electrodes – big gaps. A study is made of the characteristics of some of the discharges that could be used as auxiliary ones. It is established that the formation of a SSVD is affected significantly by the distortion of the electric field caused by the presence of a space charge of the electron flux. Dynamic profiling of electrons is found to be possible, and a SSVD is attained in systems with a strongly inhomogeneous initial distribution of the electric field in the gap. It is scalable technology!

The simplicity of the proposed methods is promising for applications in CO_2 lasers. Although the main results of the current investigation are obtained for CO_2–N_2–He gas mixtures, the methods can also clearly be applied to other gases such as, for example, N_2O. Because of our investigations, the aperture of SSVD-based lasers and amplifiers was increased up to 1m for an electric efficiency of more than 15%. The 30kJ per pulse laser had been developed and tested (Figure 18). Such an approach is suitable for realization of MJ scale laser with aperture up 2x2 m^2 due to the discharge gap scalability. That project was done on paper only.

Figure 18. 30 kJ CO_2 SSVD laser.

Just a little later, the SIVD in SF6-hydrocarbon mixtures, used as the active media for non-chain HF/DF lasers, was investigated as well. The following features of the development of the SIVD are established. After the primary electrical breakdown of the discharge gap, the SIVD spreads in the gap in the direction perpendicular to that of the electric field as a result of the consecutive appearance of overlapping diffuse channels. As the new channels appear, the current that flows through the channels formed earlier diminishes. The volume occupied by the SIVD increases almost linearly when the energy deposited in the plasma increases and, when the discharge volume is confined by a dielectric surface, the discharge voltage increases simultaneously with an increase in the current. The hypothesis is put forward that mechanisms exist for the limitation of the current in the conducting channel. They are associated with the specific energy released in the plasma, and they prevent the transfer of all deposited energy through a separate channel. It is shown that such mechanisms can be the dissociation of SF6 and the attachment of electrons to excited SF6 molecules. The main result for SIVD – based HF/DF lasers is very much promising for further scaling up, and 10-100kJ level for the pulsed energy output exceeds any possible limit for this type of laser system.

Our estimations of the space debris elimination business are promising for this type of new lasers- SIVD based HF/DF laser. Such a laser can be easily realized for the low rate P-P mode of operation – 15-20 Hz (Total power – 1, 5 MW), which is necessary for the applications under consideration – cleaning of space at the height up to 400 km. For the telescope with main mirror $D = 6$ m, wavelength of radiation -3.8 mkm (good for propagation through the atmosphere) and pulses time duration 1 mks an attainable level of intensity for 2 dif. limits – few tens of MW/cm^2. Total weight of the system can be fixed at the level of 15 tones. The size of the system -40ft container. Main advantage of the system is a non-hazardous nature of components, which means that the usage of such a system could be very much effective.

CONCLUSION

We investigated the formation of the SSVD in CO_2-N_2-He mixtures by preliminary filling of the gap with an electron flux from an auxiliary discharge plasma. This method was suitable for large distances between the electrodes. A study was made of the characteristics of some of the discharges which could be used as auxiliary ones. It was established that the formation of a SSVD was affected significantly by the distortion of the electric field due to the presence of a space charge of the electron flux. Dynamic profiling of electrons was found to be possible and a SSVD was attained in systems with a strongly inhomogeneous initial distribution of the electric field in the gap, the aperture of SSVD-based lasers and amplifiers was increased up to 1m for an electric efficiency of more than 15%. The 30 kJ per pulse laser had been developed and tested. Such an approach is suitable for realization of MJ scale laser with aperture up 2x2 m^2 due to the discharge gap scalability. That project was done on paper only.

The simplicity of the proposed methods is very promising for their applications in CO_2 lasers. Although the main results of the present investigation were obtained for CO_2-N_2-He gas mixtures, the methods can clearly be applied also to other gases like N_2O, for example. Due to our investigations the aperture of SSVD based lasers and amplifiers had been increased up to 80 cm. for the electric efficiency of more then 10%. At the same time the SIVD in SF6 hydrocarbon mixtures, used as the active media for nonchain HF lasers, has been investigated as well. The following features of the development of the SIVD have been established:

1. After the primary electrical breakdown of the discharge gap, the SIVD spreads in the gap in the direction perpendicular to that of the electric field as a result of the consecutive appearance of overlapping diffuse channels. As the new channels appear, the current flowing through the channels formed earlier diminishes.
2. The volume occupied by the SIVD increases almost linearly with increase in the energy deposited in the plasma and, when the discharge volume is confined by a dielectric surface, the discharge voltage increases simultaneously with increase in the current.
3. The hypothesis was put forward that mechanisms exist for the limitation of the current in the conducting channel. They are associated with the specific energy released in the plasma and they prevent the transfer of all the deposited energy through a separate channel. It is shown that such mechanisms may be the dissociation of SF6 and the attachment of electrons to vibrationally excited SF6 molecules.
4. Our results for SIVD – based HF/DF lasers are very much promising for further scaling up, and 10-100 kJ level for the pulsed energy output exceeds any possible limit for this type of laser system.
5. 5. Our estimations of space debris elimination business are promising for this type of new lasers- SIVD based HF/DF laser. Such a laser can be easily realized for the low rate P-P mode of operation – 15-20 Hz (Total power – 1, 5 MW), which is necessary for the applications under consideration –cleaning of space at the height up to 1000 km.

REFERENCES

[1] V. Yu. Baranov, V.M. Borisov, A.A. Vedenov, S.V. Drobyazko, V.N. Knizhnikov, A.P. Napartovich, V.G. Niz'ev, and A.P. Strel'tsov: Preprint No. 2248 (in Russian), Institute of Atomic Energy, Moscow, 1972.

[2] A. J. Palmer: "The article you requested is currently in production and will be available online at a later date", *Appl. Phys. Lett.* 25, Vol. 138, (1974).

[3] V. N. Karnyushin, A.N. Malov, and R.I. Soloukhin: Kvantovaya Elektron. (Moscow), Vol. 5, (1978), pp. 555. [*Sov. J. Quantum Electron.* 8, 319 (1978)]

[4] H. Seguin and J. Tulip: "The article you requested is currently in production and will be available online at a later date", *Appl. Phys. Lett.*, Vol. 21, (1972), pp. 414.

[5] A. F. BelyatskiT, D.B. Gurevich, M.A. Kanatenko, and I.V. Podmo-shenski, T. Pis'ma: "Zh. Tekh. Fiz.", *Sov. Tech. Phys. Lett.,* Vol. 6, (1980), pp. 73.

[6] S.C. Lin and J.I. Levatter: "X-ray preionization for electric discharge laser", *Appl. Phys. Lett.,* Vol. 34, (1979), pp. 505.

[7] A. I. PavlovskiT, V.S. Bosamykin, V.I. Karelin, and V.S. *Nikol'skii: Kvantovaya Elektron.* (Moscow), Vol. 3, (1976), pp. 601. [Sov. J. Quantum Electron. 6, 321 (1976)]

[8] A. I. Pavlovskii, V.F. Basmanov, V.S. Bosamykin, V.V. Gorokhov, V.I. Karelin, and P.B. Repin: *Abstracts of Papers presented at Second All-Union Conf. on Physics of Electric Breakdown in Gases*, Tartu (in Russian), (1984), pp. 330.

[9] R. V. Babcock, I. Liberman, and W.P. Partlov: "Volume ultraviolet preionization from bare sparks", *IEEE J. Quantum Electron.*, Vol. 12, (1976), pp. 29-34.

[10] A. V. Kozrev, Yu.D. Korolev, G.A. Mesyats, Yu.N. Novoselov, A.M. Prokhorov, V.S. Skakun, V.F. Tarasenko, and S.A. Genkin: *Kvantovaya Elektron.* (Moscow), Vol. 11, (1984), pp. 524. [Sov. J. Quantum Electron. 14, 356, (1984)]

[11] O. P. Judd: *J. Appl. Phys.*, Vol. 45, (1974), pp. 4572.

[12] R. E.Beverly III: *Appl.Phys.*, Vol. B53, (1991), pp. 187-193.

[13] V. V. Apollonov, G.G. Baitsur, A.M. Prokhorov, and K.N. Firsov: *Pis'ma Zh. Tekh. Fiz.*, Vol. 11, (1985), pp. 1262. [Sov. Tech. Phys. Lett. 11, 521 (1985)]

[14] A. K. La amme: *Rev. Sci. Instrum,* Vol. 41, (1970), pp. 1578.

[15] A. J. Beaulieu: *Appl. Phys. Lett.*, Vol. 16, (1970), pp. 504.

[16] R. Dumanchin and J. Rocca-Serra: C. R. *Acad. Sci. Ser,* Vol. 269, (1969), pp. 916.

[17] V. P. Gorelov, N.N. Minakova, and V.A. Chagin: Elektrotekh. *Promst. Ser. Elektrotekh. Mater.,* Vol. 5, (1972), pp. 20.

[18] V. V. Apollonov/*Central European Journal of Physics* 4 (2003) 708{756 731

[19] P. A. Belanger, R. Tremblay, J. Boivin, and G. Otis: *Can. J. Phys.,* Vol. 50, (1972), pp. 2753.

[20] V. V. Apollonov, A.M. Prokhorov et al: *Kvantovaya Elektron.* (Moscow), Vol. 11, (1984), pp. 1241. [Sov. J. Quantum Electron. 14, 835 (1984)]

[21] T. Y. Chang: "The article you requested is currently in production and will be available online at a later date", *Rev. Sci. Instrum.*, Vol. 44, (1973), pp. 405.

[22] V. V. Apollonov A.M. Prokhorov et al: *Kvantovaya Elektron.* (Moscow), Vol. 11, (1984), pp. 1327. [Sov. J. Quantum Electron. 14, 898 (1984)]

[23] V. V. Apollonov,A.M. Prokhorov et al: *Kvantovaya Elektron.* (Moscow), Vol. 12, (1985), pp. 5. [Sov. J. Quantum Electron. 15, 1 (1985)]

[24] V. V. Apollonov, A.I. Barchukov, A.M. Prokhorov et al: *Pis'ma Zh. Tekh. Fiz.*, Vol. 3, (1977), pp. 1073. [Sov. Tech. Phys. Lett. 3, 441 (1977)]

[25] V. V. Apollonov et al: *Pis'ma Zh. Tekh. Fiz.*, Vol. 5, (1979), pp. 1518. [Sov. Tech. Phys. Lett. 5, 643 (1979)]

[26] V. V. Apollonov A.M. Prokhorov et al: *Kvantovaya Elektron.* (Moscow), Vol. 10, (1983), pp. 1458. [Sov. J. Quantum Electron. 13, 947(1983)]

[27] A. A. Kuchinskii (Kuchinsky), B.V. Lyublin (Ljublin), V.F. Shanskii (Shansky), and V. A. Sheverev: *Proc. Sixteenth Intern. Conf. on Phenomena in Ionized Gases*, Dusseldorf,, (1983), (ed. by W. Botticker, H. Wenk, and E. Schulz-Gulde), *Contributed Papers*, Vol. 2, publ. by the Conference Committee, (undated), pp. 200.

[28] A. L. Ward: *J. Appl. Phys.,* Vol. 36, (1965), pp. 2540.

[29] A. H. von Engel: *Ionized Gases,* Clarendon Press, Oxford, (1955).

[30] I. L. Kamardin, A.A. Kuchinskii, V.A. Rodichkin, and V.F. Shanskii: Abstracts of *Papers presented at Third All-Union Conf. on Laser Optics* (in Russian), Leningrad, (1982), pp. 64.

[31] S. I. Andreev, N. M. Belousova, P.N. Dashuk, D.Yu. Zaroslov, E.A. Zobov, N.V. Karlov, G.P. Kuz'min, S.M. Nikiforov, and A.M. Prokhorov: *Kvantovaya Elektron.* (Moscow), Vol. 3, (1976), pp. 1721. [Sov. J. Quantum Electron. 6, 931 (1976)]

[32] Yu. D. Korolev, G.A. Mesyats: "Fizika Impul'snogo Proboya Gazov", *The Physics of the Pulsed Electrical Breakdown of Gases*, (1991).
[33] V. V. Apollonov, S.Yu. Kazantsev et al: *Pis'ma Zh. Tekh. Fiz.*, Vol. 22, (1996), pp. 60. [Tech. Phys. Lett. 22 1026 (1996)]
[34] V. V. Apollonov, S.Yu. Kazantsev et al: *Kvantovaya Elektron.* (Moscow), Vol. 24, (1997), pp. 213. [Quantum Electron. 27 207 (1997)]
[35] V. V. Apollonov, S.Yu. Kazantsev et al: *Kvantovaya Elektron.* (Moscow), Vol. 25, (1998), pp. 123. [Quantum Electron. 28 116 (1998)]
[36] V. V. Apollonov/*Central European Journal of Physics* 4 (2003) 708{756.
[37] V. V. Apollonov, N. Firsov, S.Yu. Kazantsev, V.F. Oreshkin: *Proc. SPIE Int. Soc. Opt. Eng.*, Vol. 374, (1998), pp. 3574.
[38] V. V. Apollonov, S.Yu. Kazantsev et al: II Mezhdunar. Konf. Po Fizike Plazmy i Plazmennym Tekhnolo-giyam, Minsk, (1997), pp. 154. [Proceedings of the Second International Conference on Plasma Physics and Plasma Technologies, Minsk, Belarus, 1997], [Minsk: National Academy of Sciences of Belarus, 1997)
[39] I. M. Bortnik: Fizicheskie Svoistva i Elektricheskaya Prochnost' Elegaza, *Energoatomizdat,* Moscow, 1998. [The Physical Properties and the Electrical Strength of Sulfur Hexa°uoride]
[40] V. V. Apollonov, S.Yu. Kazantsev et al: IX Konf. po Fizike Gazovogo Razryada, Ryazan, (1998), pp. 58. [Proceedings of the Ninth Conference on the Physics of Gas Discharges, Ryazan, 1998], [Ryazan: Ryazan Radio Engineering Academy, 1998]
[41] L. Richeboeuf, S. Pasquiers, M. Legentil, V.J. Puech: "The influence of H2 and C2H6 molecules on discharge equilibrium and F-atom production in a phototriggered HF laser using SF6", *Phys. D,* Vol. 31, (1998), pp. 373.
[42] A. F. Zapol'skii, Yushko: *Kvantovaya Elektron.* (Moscow), Vol. 6, (1979), pp. 408. [Sov. J. Quantum Electron. 9 248 (1979)]
[43] E. B. Gordon, V.I. Matyushenko, P.B. Repin, V.D. Sizov: *Khim. Fiz.,* Vol. 8, (1989), pp. 1212.
[44] A. A. Radtsig, M. Smirnov: Spravochnik po Atomnoii Molekulyarnoi Fizike (*Handbook of Atomic and Molecular Physics),* Atomizdat, Moscow, 1980.
[45] H. S. Massey: *Negative Ions*, Cambridge University Press, Cambridge, 1976.
[46] D. I. Slovetskii, A.A. Deryugin: "in Khimiya Plazmy", *The Chemistry of Plasma,* (1987), pp. 240.
[47] A. A. Belevtsev: Proceedings of the Thirteenth International Conference on Dielectric *Liquids* (ICDL'99), Nara, Japan, 1999, pp. 672.
[48] A. V, Melekhov: *Abstracts of Papers presented at Second Ail-Union Conf. on Physics of Electric Breakdown of Gases* (in Russian), Tartu, 1984, pp. 357.
[49] V.V. Apollonov, S.Yu. Kazantsev *et al: Izv.* Akad. Nauk, Ser. Fiz., Vol. 64, (2000), pp. 1439.
[50] V. V. Apollonov et al: *Proc. SPIE Int. Soc. Opt. Eng.,* Vol. 370 (1999), pp. 3886.
[51] V. V. Apollonov, A.A. Belevtsev et al: *Kvantovaya Elektron.*, Vol. 30, (2000), pp. 207.
[52] V. V. Apollonov, A.A. Belevtsev et al: Proc. XIII Intern. *Conf. on Gas Discharge and their Applications*, Vol. 1, (2000), pp. 409.
[53] V. V. Apollonov, A.A. Belevtsev et al: Proc. XXV Intern. *Conf. on Phenomena in Ionized Gases*, Vol. 1, (2001), pp. 255.

[54] N. Nakano, N. Shimura, Z.L. Petrovic, T. Makabe: "Simulations of rf glow discharges in SF6 by the relaxation continuum model: Physical structure and function of the narrow-gap reactive-ion etcher", *Phys. Rev. E*, Vol. 49, (1994), pp. 4455.

[55] V. V. Apollonov/*Central European Journal of Physics* 4 (2003) 708{756 733

[56] H. Hilmert, W.F. Schmidt: "Electron detachment from negative ions of sulphur hexa°uoride-swarm experiments", *Phys. D: Appl. Phys.*, Vol. 24, (1991), pp. 915.

[57] A.A. Belevtsev, L.M. Biberman: Izv. Akad. Nauk SSSR, *Ser. Energ. Tramp.*, Vol. 3, (1976), pp. 74.

[58] V. V. Apollonov, S.Yu. Kazantsev et al: *Kvantovaya Elektron.*, Vol. 30, (2000), pp. 483.

[59] V. V. Apollonov, A.A. Belevtsev et al: *Kvantovaya Elektron.*, Vol. 31, (2001), pp. 629.

Chapter 5

New Approach to Disk Lasers

The mono-module disk laser concept is an effective design for diode-pumped solid-state lasers, which allows the realization of lasers with a super-high output power, having very good efficiency and excellent beam quality. Since the first demonstration of the principle in 1964 the output power of mono-module disk lasers has been increased to the level of few kW in continuous wave (CW) mode. "Zig-Zag" disk laser geometry is not, in our opinion, promising for scaling up the laser output. The scaling laws for the mono-module disk laser design show that the limits for CW mode are far beyond 100 kW for the output power, and the energy can be higher than 100 J in pulsed mode. Due to efficient porous cooling technology and possibility of suppressing amplified spontaneous emission (ASE), mono-module disk lasers can operate in CW and pulse-periodic (P-P) regimes at extremely high output powers.

Introduction

It has long been customary that as new technologies emerge into the light, potential users and experts start wondering whether these new techs will replace the old, well-established approaches to the solution of known problems. When we look back at the time when the first laser was created, it becomes clear that each new and more advanced technology usually replaces the pre-existing and well-proven technologies. However, a more differentiated and balanced assessment of the many innovations in the field of laser shows that there is no such thing as a perfect design, or an ideal laser. There is always room for improvement and further development. It is possible that in the near future a laser with disk geometry of the active medium will become the dominant technology. However, despite this, a number of existing laser technologies (with some exceptions, of course) will continue to improve for quite a long time. Based on these considerations, we must continue to develop most types of lasers, each time clearly specifying their undeniable technological niche and knowing well their advantages and disadvantages.

High efficiency and excellent beam quality of disk lasers [1] mean that they can be widely used in modern science and industry for a very large range of applications, including treatment of the surface of dielectric materials in microelectronics, cutting, drilling, welding, polishing and cleaning of the surface, other technological operations with superhard and fragile metals and composite materials, etc. Pulse-periodic *Q*-switched disk lasers and high-average-power, mode-locked laser systems ensure optimal conditions for ablation (sublimation) of a material. A number of advantages of high repetition rate P-P lasers emitting short pulses, as applied to a wide range of industrial technologies, is the basis of many modern concepts of disk lasers. High-intensity light with an insignificant thermal lens effect in the central highly loaded zone of the active medium have led to the lifting of restrictions on the brightness of the pump diode. This has reduced the cost of laser sources, and thus has significantly improved the efficiency of electro-optical conversion, especially in the regime of high average powers. The power of the laser source is varied by scaling the cross sectional area of the generated radiation. It should be noted that its ratio to the thickness of the active material in disk geometry is much larger than the ratio of the cross section of typical core elements of the active medium to their length in any conventional solid-state laser system. This eliminates the problems associated with nonlinear distortions of the geometry of the active medium, and enables realizing super high peak and average powers of laser sources with disk geometry and same parameters of radiation in the far-field region.

Disk and Fiber Lasers

It is clear that each laser source has its advantages and disadvantages. The beam quality of modern solid-state lasers (disk and fiber lasers) is pretty excellent, and their efficiency is very high. A fiber laser has significant advantages at low, average and peak output powers, while a disk laser, on the contrary, is preferable in those applications which require high and super-high output powers.

Among high repetition rate P-P sources emitting coherent high average and peak output radiation, disk lasers will undoubtedly dominate today and in the future. They provide the best solution for many industrial applications in the multi-kilowatt power range. A disk is a simple and easily excited laser element, which allows cost-effective generation of radiation with improved parameters. It has a lot of potential for the future. The advantage of a disk laser over a fiber laser is already obvious in the kilowatt power range. Having a large radiating surface, the power density of the disk laser is not critical even at high peak powers. On the contrary, in fiber lasers the increase in the peak power negatively affects the reliability of the resonator. The main disadvantage of the fiber laser is also its high sensitivity to the reflected beam, which often appears in the laser due to the interaction of the generated light with matter. If the radiation reflected from the target affects the fiber laser resonator, then it should be immediately shut down to preserve the cavity.

At the same time, the disk laser resonator is insensitive to the penetration of reflected beams. Even in the case of highly reflective materials, technological operations can be performed without the risk of an emergency switching off of the laser. Another advantage of kilowatt disk lasers is their modular design, which allows one to replace laser modules during after-sales service, thereby ensuring that in the event of failure of the laser, the downtime as well as the cost of repairs will be minimal.

On the contrary, in fiber lasers, the single-unit design of the resonator makes it impossible to easily replace the failed unit without extensive alteration of the system as a whole. At the same time, the use of fiber laser technology offers certain advantages, especially in the power range (up to several hundreds of watts). This laser is suitable for micromachining of dielectric surfaces and composite materials. However, only the disk laser has a high potential in the case of high peak powers, i.e., when it is necessary to provide a train of short pulses with the highest peak power. In a typical disk laser, as opposed to a fiber laser, the power density inside the cavity is far below the critical damage threshold of the laser active medium and surfaces of auxiliary optical elements, even at a high peak power. For this reason, a disk laser allows, among other things, a better use of the laser medium in general, and significant improvement in laser efficiency.

Thus, in the future, solid-state diode-pumped disk lasers will play the dominant role. Laser diode structures are central elements in many new types of lasers. Today, the decisive factor is the acquisition of the necessary experience and skills to design most efficient new laser systems. Semiconductor-based technologies and applications of laser equipment on new physical principles are key technologies of modern times. For many years, the world has accumulated the experience needed today in this field of knowledge. The challenge is to ensure technological excellence in the application of laser systems, not only today but in the future.

DESIGN AND PHYSICAL PRINCIPLES OF DISK LASER OPERATION

The principle of the disk laser operation is based on the use of an active element in the form of a disk with a cooled surface. High cooling efficiency of the laser medium is provided by a large area of the disk surface, which is important from the point of view of the heat transfer process. Therefore, the average power in the beam can reach quite high values. It is important to note that due to effective heat transfer within the disk area there is no thermal lens effect, characteristic of 'rod' and 'slab' (optical range) geometries of the active element. In this case, cooling is performed through a side surface of the active element; a two-dimensional heat flow, forming a parabolic profile of thermal distortion, propagates through this side surface. The temperature dependence of the refractive index in this case leads to the emergence of a strong thermal lens effect. Such a lens deteriorates the directivity of the output and significantly limits the effectiveness of its action. The thin disk, in contrast, is cooled through a thin contact region on one side and generates a one-dimensional axial heat flow on the other side. As a result, the temperature gradient is distributed parallel to the laser beam, which does not result in the thermal lens effect. In practice, a thin laser element in high-power disk lasers is either connected to a porous heat sink or is cooled by forced convection [2, 3].

In both cases, the undesirable thermal lenses are eliminated by the optimum layout of structural elements of the disk laser. Integral heat transfer from the active material to the porous heat sink depends directly on the area of the pump zone and on the geometry of the heat transfer contact. At a constant temperature of the active laser element, the average output power achieved is directly proportional to the area of the pump zone and, consequently, to the cross-sectional area of the laser beam on the disk. This is a unique feature of disk lasers, which allows one to change the power density at a constant energy of the beam on the disk. This property is particularly important for multi-kilowatt sources, because it makes it possible to adjust the value of the peak output power without changing many other parameters that affect the performance of the system.

At first glance, the basic challenge is the disk thickness needed for efficient heat dissipation, because the conventional scheme involves low absorbance of the active disk material. However, the pump scheme allows one to find an elegant solution to this problem. The process of excitation in the active material of the disk can be represented in the form of multi-channel optical configuration [4, 5] consisting of a parabolic mirror and a pump deflection system. The parabolic mirror focuses a collimated beam of the laser cavity containing an active element. After partial absorption of radiation passing through the disk body, some energy of the beam is reflected from a highly reflective mirror located on the backside of the disk. As a result of multiple reflections through prisms and the parabolic mirror in general we deal with 20 iterations of radiation round-trips through the body of the disk, which ensures high efficiency of absorption of light energy of the pump beam. The optical scheme of such resonators tolerates limitations on the brightness of the pump source. Generally, pumping of a high average power disk laser does not require a beam of high optical quality. That is why both laser diodes and rather complex laser diode structures with a uniform intensity distribution can be used in this case. In terms of practical use, laser diodes with a uniform intensity distribution are more preferable due to much lower energy consumption per one watt of pump power.

For efficient processing of materials: cutting, drilling, polishing and removal of a thin film from the surface of various materials, laser pulses with a high peak power and high repetition rate are needed. Typical pulse duration of a *Q*-switched Yb:YAG disk laser lies within one microsecond. *Q*-switching is provided by means of an acousto-optic modulator. Depending on the resonator geometry it is possible to generate pulses with duration from a few hundred picoseconds to a microsecond. To date, the best results achieved with the help of P-P disk lasers having a high average power are as follows: pulse repetition rate, up to 10kHz; maximum pulse energy, 100 mJ; peak power, 200 kW; and average power, up to 1000 W. However, there are fundamental physical limitations on further scaling of the disk lasers, which we will discuss below.

LAWS FOR SCALING OF DISK LASERS

The power of disk lasers is limited not only by the power of the pump and overheating of the medium but also by the losses due to the ASE and the background radiation losses in the resonator. To avoid overheating, the size of the active medium should vary in accordance with the law of power scaling. Then, to avoid large losses caused by their exponential growth during the ASE, amplification of radiation corresponding to transverse round trips should not be large. This requires a reduction in gain *G*. Gain is determined by the reflectivity of the output mirror and disk thickness. Amplification of radiation per round trip, however, should not be substantially greater in magnitude than the radiation loss per round trip along the same optical path. The difference between the gain and losses per round trip of radiation determines the optical energy that is coupled out from the laser cavity. Reduction of the gain at this loss level requires an increase in the disk thickness. In this case, at a certain critical size, the disk becomes optically too thick and cannot be pumped above the threshold without overheating. Some features of scaling can be shown on a simple model. Let *M* be the saturation intensity of the medium. The corresponding optimal thickness of the disk can be estimated as $h \sim T/Mb$.

The corresponding optimal lateral dimensions of the disk can be represented in the form: D~T/Mb2, where *T* is a parameter of the heat load. Roughly speaking, the loss per round trip of radiation must be scaled inversely proportional to the cube root of the power required: P~ n T2/M b3.

An additional problem is the efficient delivery of pump energy. In cases with a small-signal gain, single-pass absorption of the pump is also low. It follows that the effective utilization of the pump energy is necessary for the effective operation of the laser disk. To scale the output power, the medium must be optically thin, which requires a large number of passes of the pump energy through the medium. Besides, the pump energy coupled in through the lateral side of the disk can also be a possible solution for efficient pumping.

To reduce the effects of the ASE, it was proposed to use an optical cover consisting of an undoped material on the surface of the laser disk. This cover allows the spontaneously emitted photons to escape from the active layer and prevents their resonance in the bulk of epy active material. Rays cannot be reflected from the surface, as in the case of an open disk. This allows the maximum power achievable by the disk laser to be increased by an order of magnitude. Reflection of the ASE from the disk edge should also be suppressed. This can be done through the absorbing layer at the generatrix of the disk cylinder. In the regime when epy output power is close to maximal, much of the energy is used in the ASE; therefore, the absorbing layers must also have heat-accumulating radiators. In the case of the maximum pump density of the disk laser, its efficiency is quite low: Most of the pump power is used in the ASE and is absorbed at the edges of the device. In this case, the distribution of the pump energy between several disks can significantly improve the performance of the laser system. Indeed, lasers, consisting of several modules with disk elements in a single cavity, have been repeatedly reported. One of these lasers, fabricated by the TRUMPF Group [5] – a world leader in this class of laser systems – is presented at Figure 1. Figure 2 shows a "Zig-Zag" CO_2 laser created by A.I. Barchukov at FIAN in 1970 [2]. Using this laser, we predicted a lot of new effects, which later were confirmed experimentally by the scientists involved in fabrication of high-power lasers.

Figure 1. "Zig-Zag" disk laser consisting of a series of disk modules in a single cavity (TRUMPF [5]).

Figure 2. "Zig-Zag" CO_2-laser created by A.I. Barchukov at FIAN in 1970 [2].

In the quasi-CW regime, the power can be estimated by scaling the saturation intensity with a duty cycle of the pump and by multiplying the duration of the pump by the pulse repetition rate. At moderate repetition rates (e.g., higher than 1 Hz), the maximum energy of the output pulses is approximately inversely proportional to the cube of background losses b. The undoped cover can increase the average output power by an additional order of magnitude, provided that this cover does not increase the background loss. At low pulse repetition rates (in the single-pulse regime) and sufficient pump power, there is no general limitation on the energy, but the required size of the device increases rapidly with increasing pulse energy, thereby establishing a practical limit of energy. One active element, according to the estimates, can generate an optical pulse with energy of a few thousand joules, depending on the level of internal signal loss in the disk.

Disk lasers, as well as fiber lasers, have a large ratio of the cooling surface area to the gain of the laser. However, these two different concepts of the laser design also differ by the values of the achievable peak power. The beam quality of fiber lasers is determined by properties of waveguides, the core/cladding refractive index difference, and the size of the internal diameter of the fiber that transmits light. On the other hand, the beam quality of disk lasers depends on the resonator design. With increasing size of the optical pump region (its diameter is usually a few millimeters) at a constant optical power density on the disk surface, scaling of the output power becomes possible. Adjustment of the resonator also has similar features. Self-phase modulation defines the typical limit of nonlinear amplification of ultrashort pulses. It leads to an extension of the spectral line proportional to the ratio of the effective optical path in the material with nonlinear characteristics to the effective cross-sectional area of the beam multiplied by the square of the pulse duration. An electro-optical switch in a regenerative amplifier of the disk laser makes a substantial contribution to the nonlinear characteristics of the gain, which is greater than the contribution from the disk. It must be emphasized that the values of the peak power, achieved today in single disk modules, are much smaller than the values obtained in 'rod' and 'slab' solid-state laser systems.

However, the level of the average power generated by a single disk module also leaves much to be desired.

REGENERATIVE AMPLIFICATION OF PULSES

At present, in the scientific and technical research and manufacturing processes relying on the use of high average power lasers, only sources operating in two regimes find applications, i.e., CW and P-P laser sources with a pulse repetition rate up to hundreds of Hz and pulse duration from tens of microseconds to tens of milliseconds. In most processes, purely thermal mechanism of action is implemented, since use is made of the possibility of a laser source to deliver quite a lot of energy to a small area of the surface of the work piece. High-frequency, high-average-power laser systems operating in the *Q*-switched regime, provides the pulse duration in a periodic train from a few to hundreds of nanoseconds, allows for a fundamentally different mechanism of interaction of radiation with matter, i.e., ablation, ensuring local release of energy not only in space but also in time [6]. This results in an explosive local evaporation of the substance without an intermediate liquid phase. This mechanism can significantly extend the range of technological applications of laser sources. Today, it has found a real use in practice, only at an insignificant level of output power in the range of 1000 W. However, there arises a reasonable question: Why are high-power P-P lasers not available at the international laser market? The answer is simple: The complexity of their manufacturing consists in the inability of application of the classical modulation methods to laser systems with high average and peak powers. At the same time, it is quite clear that the creation of high-power (>1 kW) P-P lasers with frequencies up to 100 kHz or higher (with a peak power exceeding the average power by two to three orders of magnitude) would significantly expand the field of application of laser sources, increase their efficiency, and implement qualitatively new effects [7].

One of the successful trends in implementation of new methods of high-frequency modulation of the output from high-power laser sources is due to the regenerative amplification of a small signal injected into the cavity of a high-power gas or solid-state laser with the classical rod geometry. Use of intracavity modulators in high-power laser systems is hindered, because high power density inside the cavity leads to the appearance of plasma on the surface elements of the modulator, to the screening of radiation and to the destruction of the optical elements of the modulator. Quite promising is the method of realization of the P-P regime in high-power lasers, which is based on the use of a self-filtering cavity. This cavity is a confocal resonator consisting of two spherical mirrors of different curvatures; in the common focal plane there is an annular output mirror with a hole. Because of the high degree of discrimination of the higher modes with respect to the resonator losses, the lower mode is reliably generated. The spherical resonator mirror is placed outside the active medium and contains a modulation unit comprising spherical mirrors. The laser mode should be formed quickly enough in the leading edge of a giant pulse. The problem of laser micromachining usually involves cutting, drilling, surface cleaning and polishing, as well as removal of thin layers due to the ablation of the material. Micromachining relies on lasers emitting pulses of different duration. The beam geometry and the pulse repetition rate are crucial: usually the pulse duration does not exceed a few tens of nanoseconds, and in some cases reaches several hundreds of picoseconds. Disk laser technology provides excellent flexibility in the scaling of the output power by combining it with a high pulse energy and excellent beam quality.

Prospects for Scaling the Power of Disk Lasers

As mentioned above, the design of the disk laser generating a continuous and relatively high power is ideal for cutting and welding metals, where high optical quality of the beam is required. It is important for such industries as automotive, transportation, aerospace and heavy engineering. However, the design of the disk laser is more suitable for a range of new technologies that are currently in demand. Today, a multi-module laser with a power up to 16 kW and beam quality of ≥2 mm mrad, fabricated by the TRUMPF Group, is the undisputed leader in this class of laser systems. CW solid-state diode-pumped disk lasers demonstrate high efficiency. The beam quality of the laser disk is outstanding, which makes it possible to work with a target from a large distance, while providing an extremely high concentration of radiation in the interaction region with the help of the focusing optics. In this design of the laser module the disk size is no more than 4−5 mm, because at larger sizes the energy loss due to the amplified spontaneous emission increases at an absolutely unacceptable rate. The above-presented laser system consists of a set of disk modules in a single cavity. This geometry of the laser system allows implementation of the P-P regime with a fairly high peak power, but the average power lies within a kilowatt. Parallel operation of disk lasers can increase the peak power of the entire system, but the phase-locking of the disk channels in the P-P regime requires additional investigation. Further expansion of the power to the megawatt level of both the average and peak power in both geometries is very problematic.

At the same time, a different approach is known to the implementation of scalable solid-state laser systems, which consists of a set of active elements in the form of 'slabs', followed by phase-locking of the generated radiation. The team of the Northrop Grumman Corporation has created a laser with a power of >100 kW and high laser beam quality equal to the diffraction limit of 1.5 (averaged value) with the exposure time of 300 seconds [8]. The laser efficiency reaches 30%. The authors of the project point to a simple replacement of individual laser channels in case of failure. They also mention some of the advantages of the parallel structure of the amplifying channels in terms of ease of further increasing output power, if necessary, to 100 kW. Beside, the laser assembled according to this scheme provides the divergence at a level of two diffraction limits from the common (composite) aperture. From general physical considerations, we can assume that for a given power level and a reasonable value of radiation resistance of optical elements, the divergence of the laser radiation at the level of 2×10-5 rad can be achieved for the CW generation regime and 0.6×10-5 rad for the P-P regime. A further increase in the number of channels in order to obtain an output power of 1 MW will require the coherent summation of the power from at least 80 channels, which seems an elusive task.

The question arises as to how the average power of a few megawatts can be achieved on the basis of solid-state lasers. And it is this power that is needed to address many problems associated with the removal of debris from near-Earth space, with launching of missiles with the help of lasers, with the creation of long-range conducting channels and others. Fiber lasers are not applicable for these purposes because of the smallness of the area of the exit pupil of the fibers and hence the impossibility of operation of such lasers in the high-frequency P-P regime with high peak powers at an average power of a few megawatts [9, 10]. The laser system based on 'slabs' also seems hard to implement because adjustment of the system and its maintenance in safe operation mode are comparable to the complexity of working with a multi-element system for the solution to the problem of controlled thermonuclear fusion (CTF) at Livermore (USA). Proceeding from the above, the answer is quite clear: Such a laser can and should be based on mono-module disk laser geometry!

This is due to the indisputable advantages of the disk geometry in terms of the minimal thermal lens in the active media and the high radiation resistance of the disk in the P-P regime because of the large area of the optical surface to couple out the radiation. Thus, the necessity to find a solution to the problem of the ASE suppression along the diameter of the disk was the major problem (matter of patent). In our case, the size of the disk at a megawatt level of the average power output should be at least 50 cm, i.e., at least a hundred times bigger than the size of the disk that is used today in existing systems. Radiation from such a laser, obtained during generation in the active medium of a single disk, does not require additional phase-locking. At the same time, such a laser in mono-module geometry will be very well combined with a large-diameter telescope for ensuring a high peak power density of the laser pulse on space debris. It is well known that the disk geometry of a laser was proposed 52 years ago [11], but only now the solution to the problem of the ASE suppression with increasing transverse dimensions of the active medium in mono-module disk geometry is found! Thus, the problems in creating new versions of the mono-module disk lasers for a new class of cutting edge applications still remains to be solved [12-14].

Conclusion

Only the future will show the total potential of mono-module disk lasers which will be used effectively for the processing of materials and for the solution of other important challenges faced by future science and technology. However, today we can say that creation of megawatt high-repetition-rate P-P disk lasers with a large cross section of the active medium will open up an avenue for their use in solving the problems of launching small satellites with lasers, formation of super-long conducting channels in space and atmosphere, cleaning of the near-Earth space from the space debris, etc.

References

[1] Basov N.G., Bogdankevich O.V., Grasiuk A.Z., *IEEE J. of, QE-2* (9) (1966).

[2] Apollonov V.V., Barchukov A.I., Prokhorov A.M., *JEEE J. of QE-6,* 10, (1974), Report of FIAN (1971).

[3] Apollonov V.V., Prokhorov A.M., Guenther A.H., *Laser Physics* V. 11, N.8 (2001).

[4] Giesen A., Hugel H., Voss A., Wittig K., Brauch U., Opower H., *Appl.Physics* B58, 365 (1994).

[5] SPIE Events Europe, TRUMPF`s R@D, (2012).

[6] Apollonov V.V., Corkum P.B., Teylor R.S., Appl.Phys.Lett., 35,147 (1979).

[7] Apollonov V.V., Kislov V.I., Kijko V.V., Suzdal`tsev A.G., QE,33 (9) (2003)
[8] SPIE Events Europe, Northrop`s R@D, (2012).
[9] Apollonov V.V., Symposium HPLS@A-2012, Istanbul 10 September (2012).
[10] Apollonov V.V., Kislov V.I., Prokhorov A.M., QE 23(5) (1996).
[11] Grasiuk A.Z., QE 42(12) (2012).
[12] Apollonov V.V, Natural science, 4, 713 (2012).
[13] Apollonov V.V., Natural science, 4, 719 (2012).

Chapter 6

High-Power/Energy Mono-Module Disk Laser Technology

A.M. Prokhorov General Physics Institute (GPI), Russian Academy of Sciences (RAS), has developed and implemented a high-repetition-rate, repetitively pulsed operating regime of high-power/energy lasers. Laboratory prototypes of gas-dynamic and solid-state lasers operating in the high-repetition-rate, repetitively pulsed regime have been designed. Successful experiments on the practical use of this regime have been performed to simulate a laser rocket engine and an orbital-scale conducting channel. Physical foundations of a laser system, which can be used to remove space debris, have been substantiated. A solution to realization of a mono-module disk laser, scaled to a level of an output power of several tens of MW, has been found. A large amount of research has been conducted in other promising areas of employment of high-power, repetitively pulsed lasers with a high pulse repetition rate. The priority projects requiring special attention include:

- a technology of 'functional' interaction of an optical pulsating discharge with an irradiated object;
- a laser engine to launch a new class of objects into space and development of a controlled interceptor;
- an orbital-scale conductive channel to design a new class of special equipment and to deliver energy from space to Earth and from Earth aboard a space station;
- a system to remove space debris and hazardous objects of natural origin; and
- a high-power, high-repetition-rate, repetitively pulsed disk laser.

Description of the Problem

Since 1964, when Russian scientists N.G. Basov and A.M. Prokhorov were awarded the Nobel Prize in Physics, our country has become a world leader in the field of laser physics and laser technology. Along with space and nuclear technologies, laser technology represents a solid foundation for a successful elimination of the technological backwardness of other industries. Russia can and should speed up work on a new generation of laser systems, physical and technical principles of which have been developed and are still preserved, despite very limited funding, by a number of the country's institutes. Everything that has been laid down by the founding fathers of Laser Physics (N.G. Basov, A.M. Prokhorov and D.F. Ustinov) in previous years is still developing. We are still the world leader in a number of laser technologies and are invited to the most important world forums. China, Brazil, South Korea and India, being outsiders in this field of knowledge, rushed however after the US, Germany, England and Japan and began to actively invest in the development of laser equipment and technology, which are the basis of high-precision instruments. The state of affairs in the field of financing of new developments and the real level of Russian experimental facilities seems today at odds with the status of a great power, because serious political and economic processes of the 1990s will affect the course of development of the laser industry of our country for a long time. What we need now is considerable intensification of research and expansion of modern laser equipment manufacturing industry. Russia continues to be a world leader, but we need to change our attitude to laser technology, which essentially strengthens the defense capability of our country.

Here we present a project related to the development of strategic technology of a scalable, repetitively pulsed mono-module disk laser. Mono-module disk laser geometry suits best for realization of a compact and lightweight laser, which, at an average output power of ~25 MW, can be placed aboard existing aircraft. The already reached ratio for solid-state laser systems allows one to use on-board mono-module disk laser systems with an average power mentioned above and to solve problems inherited from the Strategic Defense Initiative.

A series of priority technologies include:

- near-Earth space ecology, i.e., removal of dangerous space debris and objects of natural origin;
- laser engine to launch a new class of objects in space and to develop controlled interceptors;
- orbital-scale laser channel to create a new class of geophysical equipment and energy delivery from space to Earth and from Earth aboard the Space Station (long range shocker); and
- military and industrial applications based on the sublimation of a material in the zone of interaction of radiation with matter and the functional impact of an optical pulsating discharge on an object being irradiated.

This project is based on a completely new generation regime of high-power laser radiation, i.e., repetitively pulsed regime with a high pulse repetition rate (>10 kHz), proposed and developed by us. At regularly scheduled workshops on this issue – ISBEP and HPLS@A-2012 (Istanbul) – a clear conclusion was drawn that it is needed to develop high-repetition-rate, repetitively pulsed lasers generating the highest possible specific propulsion of laser engines, exceeding the best performance of all other possible types of engines. The high efficiency of the use of high-repetition-rate, repetitively pulsed regime to solve the problem of effective destruction of space debris has been shown. Leading experts in the field of high-power, high-repetition-rate, high-power/energy lasers work at the Energomashtehnika Ltd., founded at the GPI RAS by Academician Prokhorov in the difficult years of the early 1990s. We have proposed and experimentally realized a laser engine based on the mechanism of high-repetition-rate optical pulsating discharge and received record-high performance of the engine. On the basis of a high-repetition-rate, repetitively pulsed laser and related optical pulsating discharge, we have proposed and implemented a conducting channel with a minimum specific resistance and have shown the possibility of its scaling up (to the orbital scales) as well as physical feasibility of highly conductive channels in a vacuum.

In today's world, the topic of laser cleaning of the near-Earth space from space debris is being actively discussed. Thus, the space technology of debris removal by long-wavelength, low-repetition-rate laser systems proposed by scientists from the US reached its limit in the late 1990s (Strategic Defense Initiative) and proved ineffectiveness both in interaction with objects being irradiated and during transport of radiation to the object.

Thus, the renaissance of the laser launch and related new technologies of extra-long conducting channels, space cleaning and functional impact on objects can be successful only if use is made of high-repetition-rate, repetitively pulsed laser systems, which have already proved their high efficiency. High scientific culture, international competition in the field of high technologies, presence of both historical and newly established fundamental scientific resources in the Russian Federation in this segment allow one to find with a high probability several megaprojects, which can form the basis to continue the development of both laser physics and related technologies as well as scientific and educational processes in order to foster a new cluster of highly skilled professionals.

PROPOSED SOLUTION

We have developed new high-tech products of laser physics and quantum electronics:

- high-repetition-rate, repetitively pulsed operating regime of high-power/energy laser systems, initially operated in CW regime (gas-discharge lasers and diode-pumped solid-state lasers);
- scalable mono-module disk laser operating in CW and repetitively pulsed regimes with a high pulse repetition rate; and
- brand new range of interaction regimes on the basis of an optical pulsating discharge and, accordingly, new laser technologies that are unrealizable in the case of previously developed high-power/energy laser systems.

HIGH EXPERT ESTIMATES OF THE NEW LASING REGIME HAVE BEEN GIVEN BY THE FOLLOWING SPECIALISTS

Professor A. Hill, prominent researcher of the Institute of Quantum Physics at the University of Texas (the USA); A. Watanabe, president of Beams Inc. (Japan); V.V. Tsibliev, head of the Yuri Gagarin State Scientific Research-and-Testing Cosmonaut Training Center; G.A. Popov, corresponding member of the RAS, head of the section 'Fundamental Problems of Implementation of Space Flight' of the Space Council of the RAS, director of the Moscow Aviation Institute (National Research University); A. Kantrowitz, Professor of Engineering at Dartmouth College (the USA); L. Mirabo, Professor of the Polytechnic University (the USA); B.M. Belousov, former Minister of Defense Industry of the USSR; and

A.P. Sitnov and A.A. Rakhmanov, former chiefs of Armaments of Russia's Armed Forces. Our report at the meeting of the Scientific and Technical Council of the Russian Federation on new laser technology was supported by Academician N.P. Laverov and Academician B.I. Katorgin.

EXPECTED RESULTS

The development of new technologies and training of specialists in innovative fields of knowledge, such as:

- laser launching of small satellites without a carrier rocket, creation of an effective interceptor of debris in the surrounding space;
- development of a new generation of laser systems, i.e., high-repetition-rate, repetitively pulsed lasers operating in the regime of material sublimation, which are capable of processing extra-hard and brittle materials at room temperature, and creation of new technologies on this basis;
- development and creation of high-power, high-repetition-rate, repetitively pulsed laser systems for purification of water surfaces from oil products;
- development of laser systems for next-generation lithography at a wavelength of 13.5 nm or less;
- transmission of energy over long distances via a laser-plasma channel;
- creation of solid-state UV laser systems without nonlinear conversion of harmonics;
- development and creation of highly efficient, high-power, diode pumped mono-module disk lasers;
- generation of sound, light and electromagnetic fields on the basis of high-repetition-rate optical pulsating discharges;
- development and creation of long-range shockers to fight terrorists;
- development and creation of a super-stable power (static or adaptive) and large astronomical optics based on silicon carbide;
- development and creation of high-power/energy, high-repetition-rate, repetitively pulsed gas and solid-state lasers for wide application in industry and research;
- development of new technologies necessary for the implementation of high-repetition-rate, repetitively pulsed regime in high-power/energy laser sources; and
- troposphere ecology and control of extreme natural processes, as well as natural and man-made processes in the global energy supply chain of the planet.

ADDITIONAL EXPECTED RESULTS WITHIN THE FRAMEWORK OF THE PROJECT

- reduction of surface friction and increase in wear resistance of metal products using radiation of repetitively pulsed lasers;

- reduction of viscosity of gas and liquid media using an optical pulsating discharge in the interests of aviation and submarines;
- employment of laser techniques to increase the efficiency of ice-breaker fleet for offshore Arctic development projects and travelling through heavy ice cover in the Arctic Ocean;
- highly effective cleaning of surfaces of engineering, technical and architectural structures, as well as paintings and objects of art with micron precision and high speed;
- development of systems for the effective investigation of the electromagnetic compatibility of devices in confined and cramped volumes; and
- development of systems generating high-frequency sound, light and electromagnetic fields for a wide range of applications.

The products developed in the framework of the project can help solve urgent technical problems in a number of industries. Thus, the leading companies of oil industry, manufacturers of building materials (glass production), car manufacturers, companies involved in waste disposal, space industry, etc. will be potential consumers of the products and will actively use them because the proposed solutions are not only unique and modern, but also less expensive. For example, high-repetition-rate, repetitively pulsed laser systems offer numerous advantages and are capable of saving billions of dollars when it comes to cutting complex multilayer glass packages in the construction industry, as well as in optical and automotive industries for cutting curved glass and pre-shaping of objects of complex shape.

Implementation of the project:

- is sectoral, i.e., forms a basis for new branches of industry, which involves additional opportunities for modernization of the domestic economy and the formation of new, high-tech areas of production;
- promotes the creation of new jobs;
- enhances the international prestige of Russia, allowing us to make a breakthrough in leading industries (space exploration, communication, oil production, construction, etc.), because the presented technology and the products developed on its basis are unique and have no analogues in the world; and
- provides next-generation laser systems, which, due to short duration and high repetition rate of pulses at a high peak power allow for a transition to a more efficient regime of material sublimation instead of melting.

Design and Development of High-Power/Energy, Repetitively Pulsed Mono-Module disk Lasers

A high-power/energy, repetitively pulsed mono-module disk laser is a revolutionary step in the development of physical and technical foundations and technology of disk lasers. It opens up new prospects for their further development and effective solutions of a new class of both civilian and military problems. Diode-pumped solid-state disk lasers are highly efficient.

The beam quality of a disk laser is outstanding, which allows long-range irradiation of an object, while providing an extremely high radiation density in the interaction region by using focusing optics. Usually, the size of the disk is no more than 10 mm, because when the disk is large, the energy losses due to amplification of spontaneous emission outside a laser channel increase with an absolutely unacceptable rate. The above-presented laser system consists of a set of disk elements in a single cavity. This geometry ensures the repetitively pulsed regime with a fairly high peak power; however, the average power in this case is in the range of a few kilowatts. Parallel operation of disk lasers can lead to an increase in the peak power of the system, but phasing of disk channels in the pulsed regime requires additional scientific research. At present, TRUMPF has produced a laser with a virtually limiting power of up to 26 kW with a beam quality of ≥ 2 mm • mrad. A further increase to the desired multi-megawatt level of average and peak powers in both geometries is problematic. Most likely, the fate of this laser will be the same as that of quasi-optical CO_2 lasers, which were very popular in the early 1970s. The photo below shows the laser created by A.I. Barchukov with co-workers (GPI) in 1970. Using this laser we were able to anticipate many of the effects, which we subsequently had to face in designing megawatt lasers: for example, power and adaptive optics, exceptional role of silicon carbide for production of optical elements for high-power lasers and astronomical telescopes, non-crucible crystal growth, and many others.

Another approach to scalable solid-state laser systems relies on the use of a set of active elements in the form of plates (slabs) with subsequent phasing of generated laser radiation. Thus, Northrop Grumman has developed a 100-kW laser with a high laser beam quality, which is equal to 1.5 of the diffraction limit (average value) at a time of exposure of at least 300 s. The laser efficiency is 30%. A 100-kW laser assembled by this scheme provides a divergence at a level of two diffraction limits of the total (integral) aperture. A further increase in the number of channels to an output of 1 MW will require parallel summation of power from at least 70 channels, which is very problematic. The question arises: What laser should we use to provide an average power of several MW and small dimensions and weight and what power is required to solve many problems associated with the removal of debris from the near-Earth space, launching of laser-energized rockets, creation of long-range conducting channels, etc. Fiber lasers are not applicable for this purpose because of a small cross-sectional area of the exit pupil of fiber and hence the impossibility to operate such lasers in the high repetitively pulsed, high peak power regime at an average power of a few MW. Gas-dynamic and chemical lasers, due to the low efficiency and bulkiness, also seem inappropriate. Therefore, the answer is quite clear, such a laser today and for a long time in advance is a mono-module disk laser!

The choice of such a laser is dictated by the indisputable advantages of disk geometry in terms of the minimal thermal lens in the active substance and high radiation resistance of the disk in the repetitively pulsed regime because of the large area of the optical surface from which radiation is coupled out. However, it is necessary in this case to find a solution to the problem associated with the suppression of amplified spontaneous emission along the diameter of the disk. The size of the disk at a multi-megawatt level should be no less than 500 mm, which is much larger than any disk that is used today in existing systems. Radiation from the laser system being developed – ‘mono-module disk’ – does not require additional phasing. At the same time, such a laser in a single module will be combined very well with a large-diameter telescope to provide a required high-peak power density of a laser pulse on space debris and to address a number of other tasks in the interest of the Defense Industry. An alkali metal vapor laser proposed at the Lawrence Livermore National Laboratory (USA) is not a panacea and many laboratories in the world are already disillusioned with it. Its improvement will require the solution of many problems that are already openly discussed at conferences. However, it is clear that due to a much lower energy density in alkali metal vapor than in a solid, which relates directly to the dimensions and weight of the whole system, such a laser does not have a significant advantage in comparison with known chemical laser systems. Moreover, fast pumping of alkali metal vapor through the resonator suffers from the same well-known problems, leading to poor quality of output radiation.

Fifty-two years have passed since Academician N.G. Basov first proposed the disk geometry of a laser. Today, we have finally solved the problems of amplified spontaneous emission suppression in the disk geometry and of efficient cooling of large diameter disks! Thus, the prospect of a 'superlaser' for a new class of problems is at hand! A mono-module, scalable large-diameter-disk laser can and should be made by us in the near future. This will allow Russia to re-take the lead in this fundamental direction of laser physics. Such a laser can help solve a new class of tasks described in this project. The 'mono-nodule disk laser' program very expensive, but its implementation can bring fabulous dividends to creators. Assessment of the cost and timing will depend on the laser output parameters, as the lion's share of funds will be spent on the arrays of laser diodes to pump the disk. Scalability can be proved with the disk size in the range 100 – 150 mm. In this case, a disk of this diameter will provide power of more than 100 kW! Accordingly, at a disk size of 500 mm the laser output can be up to ten megawatts. It is important to note that the technology of production of almost all elements of a mono-module disk laser has been developed within the framework of other programs and we only need a refinement stage to combine all the elements of the technology into a single unit to achieve the final result.

FABRICATION OF ACTIVE ELEMENTS OF SUPER-HIGH-POWER/ENERGY LASER SYSTEMS FROM SINGLE CRYSTALS AND CERAMIC MATERIALS BASED ON YTTERBIUM-DOPED CALCIUM FLUORIDE

One of the main problems of quantum electronics is the development of technology of high-power/energy, diode-pumped solid-state lasers and organization of industrial production. The most important parts of a high-power laser system are the active elements of high optical quality and large size. The active elements can be made of laser glasses, laser crystals and optical laser ceramics. Currently, the best results have been obtained using active elements made of oxide ceramics doped with Nd^{3+} and Yb^{3+} ions. Industrial production has been organized to manufacture ceramic YAG elements with high optical transparency and larger sizes than the best single-crystal material. Ceramic YAG elements produced by sintering pressed nanopowders provide better control of the activator distribution than that in growing single crystals. There may be higher concentrations of the activator without the formation of clusters and tensions that develop during the crystal growth from the melt.

In connection with the development of diode pumping, ytterbium ions (Yb^{3+}) start playing an important role as an activator, which is due to the smaller value of the so-called quantum effect, i.e., the less difference between the pump wavelength (974 – 988 nm) and the lasing wavelength (1030 – 1050 nm for Yb compared to 1080 nm for Nd). Using the proposed combination leads to smaller heat losses in the generation of laser radiation. The second circumstance contributing to the popularity of ytterbium lasers is the absence of concentration quenching due to creation of media with a high concentration of the activator. This circumstance has led to the search for suitable matrices doped with ytterbium. Of oxide matrices, the most successful matrix to increase the output power of ytterbium lasers has always been fused silica. Silica doped with ytterbium is used to manufacture low-loss optical fibers. The output power of fiber lasers exceeded today 100 kW.

However, a more suitable matrix for Yb^{3+} ions is a fluoride matrix as it has a very significant advantage over other materials. This class of materials exhibits less pronounced nonlinear effects that lead to losses at a high beam power. Ytterbium doping levels can be raised without strong concentration quenching. Of fluoride matrices doped with ytterbium, the first place is occupied (in terms of physical and chemical properties and process ability) by $CaF_2:Yb^{3+}$ crystals. The concentration of ytterbium in them is 1.5% and 3% and the maximum dimensions constitute 100 ÷ 150 mm. These crystals are used in 'Penelopa' and 'Polaris' projects.

PREPARATION OF SINGLE CRYSTALS OF CALCIUM FLUORIDE

The formulated problem should be divided into two stages:

Stage No. 1.

Preparation and research of prototypes with a diameter up to 120 mm. This stage is of exploratory nature and aimed at studying the optical calcium fluoride crystals doped with Yb^{3+} and determining optimum characteristics and control methods of these characteristics. Namely, it is needed

- to determine the optimum concentration of ytterbium ions in calcium fluoride crystals;
- to determine the parameters, types of instruments and methods of control of the characteristics of the obtained samples. (For these tasks to be realized, we should determine (i) optical properties, spectral characteristics (purchase and calibration of a VUV spectrometer); (ii) orientation relative to the (111) plane (purchase and calibration of the diffractometer); (iii) the degree of birefringence and optical voltage (PKS-250 and PKS-125 polariscopes); and (iv) the degree of scattering of light (visual inspection). In addition it is necessary to qualitatively assess (i) the optical homogeneity (grain boundaries, slip plane, dislocation, micro inhomogeneities) using PKS-250 and PKS-125 polariscopes and (ii) the bubbles and inclusions (visual inspection under side illumination).); and
- to test the possibility of generation in the prototypes and to select the best samples in order to pass to the second stage.

Stage No. 2.

Preparation and research of prototypes with a diameter up to 250 mm. This stage is of practical nature and is associated with fabrication of large optical calcium fluoride crystals doped with Yb^{3+}. This stage also includes certifications of the obtained samples by the methods developed during the first stage. To carry out works, we must design, develop, fabricate and install a new type of vacuum-growth facilities of increased dimensions. This stage includes the development of the technology of growing highly homogeneous single crystals, the production of a pilot batch of large disks, i.e., active elements for lasers with high average and peak power. The technology being developed should be based on domestic developments and contain technological approaches approved at the first stage. This stage involves: (i) design and manufacture of non-standard equipment for the fabrication of large single crystals and (ii) development of technology for growing large, highly homogeneous single crystals.

As a result of these works we will develop a technology for manufacturing crystalline disks as active elements of lasers with high average and peak power under high-repetition-rate, repetitively pulsed operation.

FLUORIDE CERAMICS FOR ACTIVE ELEMENTS OF LASERS

Previous studies allow us to single out two main laser materials based on fluoride ceramics: $Ca_{0.95}Yb_{0.05}F_{2.05}$ and $Ca_{0.60}Sr_{0.35}Yb_{0.05}F_{2.05}$. The developed laboratory technology allows fabrication of samples with a diameter of 35 mm. The results of the research can be used as a foundation for further research, design and development of technological solutions aimed at (i) development of industrial technology for fabrication of optical ceramics based on complex fluoride chemical composition; (ii) development of technology for production of large thermostable active elements of laser systems; and (iii) fabrication of prototypes of super and ultra-compact laser systems.

Active elements made of $Ca_{0.60}Sr_{0.35}Yb_{0.05}F_{2.05}$ have excellent lasing characteristics when using diode pumping. However, the complication of the chemical composition has led to a significant drop in thermal conductivity. Therefore, such materials are not suitable for the development of high-average-power laser systems. A probable ‘ecological niche’ of these materials is low power systems, i.e., rangefinders, target designators and others. The thermal conductivity of $Ca_{0.95}Yb_{0.05}F_{2.05}$ ceramics is much higher, especially at low temperatures. It is also possible to achieve a significant increase in thermal conductivity with decreasing concentration of ytterbium in a $Ca_{1-x}Yb_xF_{2+x}$ solid solution. The results obtained are promising for the fabrication of large-sized ceramic samples for experimental works by the ‘Impulsar’ program. It is needed to increase the aperture of laser elements up to 300 – 500 mm while increasing the thermal conductivity of the disk material. A promising direction for future developments is the optimization of spectral and lasing characteristics and of thermal conductivity of the material with regard to the content of ytterbium in a $Ca_{1-x}Yb_xF_{2+x}$ solid solution. We plan to develop synthesis technology and release an experimental batch of fluoride ceramics in the form of disks for active elements of lasers with high average and peak power. In addition, we will manufacture a pilot batch of fluoride ceramic samples and organize industrial sites for synthesis and machining of blanks of fluoride laser ceramics.

Pilot and industrial production of fluoride laser ceramics must ensure the fabrication of samples in the form of 50 – 100 mm disks with corresponding characteristics. Technology of machining of fluoride laser ceramics should provide high requirements for the accuracy and cleanliness of processed elements. The production site of synthesis of fluoride laser ceramics must provide the output of samples with given specifications in the amount of not less than 8 pcs per month. The technologies being developed should be fully based on domestic developments and technological approaches.

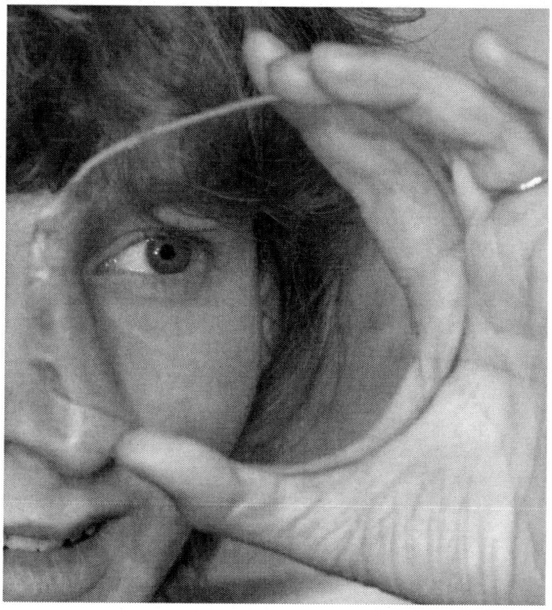

Figure 1. $CaF_2:Yb^{3+}$ laser ceramics prepared by hot-forming.

Samples of fluoride ceramics should have the following characteristics:

- microhardness, >120% of single crystals;
- ratio of fracture toughness, K_{IC} > 200% of single crystals;
- thermal conductivity, > 95% of single crystals;
- differential lasing efficiency (for laser ceramics), >35%; and
- density, 100% of the theoretical value.

The main advantage of fluoride ceramics is high mechanical strength and thermal conductivity compared to phosphate glasses and $CaF_2:Yb$ crystals.

We plan (i) to produce non-standard equipment for the manufacture of doped fluorides of metals and equipment for hot vacuum pressing (powerful press, matrix and punches of high-temperature alloys required for sintering); (ii) to develop an engineering project to produce samples of fluoride laser ceramics; (iii) to provide necessary design and technological documentation for prototypes of fluoride laser ceramics; (iv) to develop programs and methods of preliminary and acceptance tests of prototypes of fluoride laser ceramics; and (v) to fabricate samples of fluoride laser ceramics and perform state acceptance tests.

First Stage

Disk blanks with Ø30 mm and different concentrations of ytterbium. To implement the above said, it is necessary:

Stage	Costs	Period
Purchase of materials CaF2, 200 – 300 kg YbF3, 15 – 20 kg PbF2, 30 kg CF4, 300 kg		1.5 months
Preparation of initial materials		2 months
Manufacture and preparation of graphite tooling		2 months
Design and manufacture of non-standard equipment for the fabrication of large single crystals		6 months
Commissioning works, thermal testing and trial growth		6 months
Preparation of prototypes, growth of active elements		3.5 months
Fabrication and machining of active elements		1.5 months
Evaluation of characteristics		1 month

No.	Stage	Costs	Period
1	Growing of CaF_2:Yb crystals, 0.3 kg for each concentration of ytterbium		0.5 months
2	Preparation of initial materials (purchase of a ball grinder and graphite tooling for annealing in vacuum)		1 month
3	Hot uniaxial pressing under different conditions, pre-machining of materials		1 month
4	Annealing in carbon tetrafluoride atmosphere (purchase of graphite tooling for annealing in CF_4)		1 month
5	Machining of ceramic blanks		1 month
6	Study of optical and laser characteristics (equipment for studying single crystals)		1 month

Second Stage

Disk blanks with Ø100 mm and selected concentration of ytterbium. Stages of work:

1. Growing of CaF2:Yb crystals, 0.3 kg
2. Preparation of initial materials (purchase of graphite tooling for annealing in vacuum)

3. Hot uniaxial pressing under different conditions (purchase of a mold made of a heat resistant molybdenum alloy)
4. Annealing in carbon tetrafluoride atmosphere
5. Study of optical and laser characteristics (equipment for studying single crystals)

The above-described works will form the basis for the development of technological production of fluoride materials, which can be used as active elements (disks) of high-average- and peak-power lasers operating in the high-repetition-rate, repetitively pulsed regime.

Description of a High Power/Energy Laser

High-power/energy lasers should exhibit high beam quality, high efficiency and high power density on the sample; in addition, they should be compact. The main areas of application of such lasers are surface treatment, welding and cutting of materials. Today, many car factories in Germany and Japan use laser welding and cutting. The pulsed operating regime of lasers represents the greatest interest for medical and scientific applications. The main problem of a high-power laser is the heat release in the active element, which leads to strong degradation of laser characteristics. This problem is solved efficiently in fiber and disk lasers. Fiber lasers exhibit high performance, but operate only in CW regime, the maximum capacity without impairing the beam quality being ~2 KW. Disk lasers can operate in the repetitively pulsed regime and have no restrictions on the quality of the beam with increasing laser power in the above-specified ranges. Industrial production of high-power lasers is poorly developed despite the fact that research in the field of disk lasers and fiber lasers is very active, but constrained by the lack of understanding of the problem and funding. Leaders in the manufacture of disk lasers with 'zig-zag' geometry are the German companies 'Trumpf' and 'Rofin'. However, the chosen method for increasing the output power of the system by replicating the active medium in the form of small-diameter disks in a cavity is limited by the threshold values of the power density on each disk member of the active medium and requires further research for ways to increase the disk diameter.

The second direction of the project is the development of a model range of technological disk lasers with an output power from 10 kW to 100 kW and different operating regimes (continuous wave, repetitively pulsed, mode locked) for optimal performance in a variety of industrial processes. For the developed laser system to be commercialized, works should be carried out to integrate the system into a single unit. Especially interesting is the testing of domestic diode pumping, resulting in a substantial decrease in the price of the finished laser system compared with foreign counterparts.

Idea of the Project

This part of the project is aimed at developing technologies for designing a high-average-power, high-repetition-rate, repetitively pulsed mono-module disk laser. The project is based on the results of research carried out by the staff of GPI since 1985. There are three main technologies that are used in the construction of disk lasers. First, it is the use of diode pumping. The appearance of this type of pumping in the mid-1990s made it possible to improve the disk laser concept and to pass from pumping the disk by a solid-state laser to more efficient diode pumping. First, modern diode pumping is characterized by highly reliability (with a lifetime of up to 100 thousand hours) and high efficiency with respect to the consumed electric power (85%). Second, oxide laser ceramics doped with Nd^{3+} and Yb^{3+} ions are used as active elements. These dopants have a small quantum defect and no channel up-conversion (and hence low heat release during the absorption of the pump). Third, the active element of the laser acts as an active mirror – a thin disk soldered by one of the ends on a highly porous heat sink. A dielectric mirror, which is one of the resonator mirrors, is preliminarily deposited on this end. This technology and some additional fundamental solutions not only allow effective heat release from the active element, but also scaling of laser output parameters, and in particular, the main parameters of the laser system being developed, i.e., the generated radiation power due to suppression of the effect of amplified spontaneous emission. The above-described technology provides a laser power of up to 100 kW in a multimode beam. The main ideas of the project are currently described in an internal document and not patented because of the possibility of information leakage. At present, using one disk element, we managed to reach an output power of 4 kW with an efficiency of more than 40%. By increasing pump power and pump beam diameter in the active element (using the so-called scaling), we can significantly increase the output power level while maintaining the efficiency of the laser system. In addition, it is necessary to produce an active element with the help of the new technology and to optimize the dielectric and metal deposition on the active element.

What is needed for any scientific body to solve this problem? The answer is as follows:

1. The creation of technology for production of large-diameter disks.
2. The solution to the problem of suppression of amplified spontaneous noise.
3. The implementation of effective systems for disk pumping and cooling.

In addition, there arises the need to purchase several pieces of unique equipment. Namely,

1. Production of a large-diameter single-crystal disk (>100 mm) in order to reach > 100-kW output powers (in a series of disks).
2. Production of a large-diameter ceramic disk (>100 mm) in order to reach > 100-kW output powers (in a series of disks).
3. Fabrication of a sandwich from the two above-mentioned disks.
4. Production of a series of larger-diameter mirrors and a telescope with a controllable mobile base.
5. Deposition of complex coatings on the elements of the resonator and an active element.
6. Purchase of 500 kW of LD arrays for pumping active media of the disk in order to reach 40% efficiency and covering possible accidents during soldering and testing.
7. Power sources to pump LD arrays (weight of 350 kg per 1 MW).
8. Production of capillary-porous systems for cooling laser diode arrays.
9. Integrated system of disk cooling (heat flux up to 8.5 kW/cm$^{2)}$.

10. Mechanical works.
11. Optical works.
12. Assemblage of the system.
13. Devices for measuring such parameters as power, timing and divergence.
14. Electronic systems of synchronization of the pumping.
15. Purchase of large-diameter isostatic equipment.
16. Purchase of a fiber laser for the regime of radiation regeneration.
17. Purchase of equipment for ion polishing.
18. Purchase of high-purity materials for pressing and growing a disk.

Chapter 7

PLANS AND REALITY

Plans are changing for laser weapons development following three crucial demonstrations. Early in 2009, an LD-pumped solid-state slab laser from Northrop Grumman emitted 100 kW continuously for five minutes. Also, another group (Textron) exceeded the average power of 100 kW using a much different LD-pumped solid-state laser. The same month, a megawatt-class Airborne Laser (ABL) successfully completed the long-shot shoot-down of a liquid-fueled ballistic missile, representative of potential enemy military threats.

On the surface, all three demonstrations sound like welcome good news for military applications. The ABL became the first laser to destroy a ballistic missile in flight, a fundamental key milestone for missile defense. Yet the fiscal 2011 budget put the brakes on further development of the chemical oxygen/iodine laser (COIL) technology used in the ABL, while including money to develop new laser technology and convert the 747-mounted laser into a testbed to study high-energy laser effects and ways to get much higher extracted efficiency. In contrast, the budget does include plans for field trials to test potential of electrically powered 100-kW class lasers for battlefield use against rockets, mortars, artillery, and small boats. The ABL is the latest in a series of laser missile-defense projects that began in the Carter Administration. The Defense Advanced Research Projects Agency's (DARPA) Space Laser Triad included construction of a 5 MW hydrogen-fluoride (HF) chemical laser on the ground, called Alpha, to show feasibility of orbiting laser battle stations to block a massive nuclear attack. Ronald Reagan's Strategic Defense Initiative (SDI) expanded on the idea, proposing a layered defense and alternatives including pop-up bomb-driven x-ray lasers. Alpha eventually reached megawatt-class power in 1991, but the end of the Cold War and formidable problems with laser and space technology led to shifts in plans.

The ABL program was born in 1993 to defend against a more manageable missile threat — launch of a handful of missiles by any possible enemy. The plan was to install a megawatt laser in a plane that could fly close enough to potential launch sites that it could destroy missiles during their vulnerable boost phase. Developers hoped that use of a 1.3 μm COIL would help overcome beam-propagation problems, as well as allow smaller beam-direction optics than needed for HF/DF chemical lasers.

In 1996, the Pentagon issued Boeing a $1.1 billion contract to build the ABL, with lethality tests starting in the early 2000s. The Bush Administration designated the ABL as the boost phase element of a multilayer missile defense system, but the planned missile shoot-down fell years behind schedule and ran billions of dollars over budget.

The USA administration pulled the plug on plans to build a fleet of ABLs before completing tests. The administration told Congress that the laser's projected lethal range was only 85 miles (135 km), well short of the 200 km sought-for boost-phase missile defense. Next year's proposed budget calls for converting the ABL into a testbed to study laser effects in the air and on the ground. A new "Directed Energy Research" budget line will split money between ABL tests and development of new types of lasers.

The MDA did not disclose the laser's range in its last tests, but it had planned to start with close-in targets and move to more distant targets and more difficult missions. The first successful shoot-down was of a sounding rocket. The ABL shot down the rocket, locking onto it within seconds after launch, then measuring atmospheric distortion and compensating for it before firing the COIL, which caused structural failure of the missile within two minutes of launch. Within an hour, the ABL tracked and engaged a boosting solid rocket, showing it could be fired repeatedly. The second missile was not destroyed because of a "beam misalignment," according to Aviation Week Journal article. The MDA plans further tests into the frame over the next two years, which may include countermeasures and multiple missiles to assess system response. Important issues yet to be addressed in public include the laser's ability to operate at full power in flight and how well the non-cooled light optics can withstand full-power operation.

DIODE-PUMPED ALKALI LASERS ARE THE FUTURE FOR THE LASER WEAPON

The last few years analysis and experience identifies LD-pumped alkali lasers as a leading candidate for missile defense. Their advantages include thermal management in a vapor phase rather than a solid, potentially extremely high optical to optical conversion efficiency, a good match to near-infrared pump diodes, very high power per unit volume, and running on electrical power instead of requiring special chemical fuels, which are good for outer space only.

The sophisticated diodes with a very narrow line pump an absorption line of the neutral-atom $^2S_{1/2}$ ground state, exciting it to the $^2P_{3/2}$ state. Collisions with a buffer gas relaxes that state rapidly to the lowest excited state, $^2P_{1/2}$, which is the upper laser level. Intense pumping by a diode array produces a population inversion with extremely high small-signal gain.

The photon defects are remarkably small, only 4.7% in cesium, 1.9% in rubidium, and 0.44% in potassium; therefore, the optical-to-optical slope efficiency can in theory exceed 95%. Careful engineering is needed to approach those levels. At a low pressure, alkali metals have a very narrow line width, and so buffer gases are added to match the absorption band with the 2 nm emission band of the diode array. Experimental slope efficiencies have exceeded 80%. With radiative lifetimes of only 30 ns, they do not count as energy-storage media in the normal sense. "This is a quasi-two-level laser. The more you pump, the less ground state atoms there are to absorb, so you have to stimulate emission to get atoms back to the ground." This feature is not in a favor of high peak powers of pulses in the high-repetition-rate regime of lasing.

Experiments are still in the early stages, but published power levels have already reached very impressive results in static systems. Most early experiments were with longitudinal pumping, but some experiments showed that transverse pumping is promising for higher powers. In addition to a larger pump area, transverse pumping geometrically separates the closely spaced pump and emission wavelengths, which are hard to separate when they are aligned along the laser axis.

Waveguide structures might allow a multi-kilowatt output, but high velocity flowing systems in general promise a much higher power by ruling out heat and tactical systems. New engineering techniques are needed to take advantage of the unusual and very chemically aggressive system and it attracts the attention of scientists around the globe towards that type of system. Weight and sizing problems are not in favor of scaling up tendencies.

DIODE-PUMPED SOLID-STATE LASERS

Two previous demonstrations have put LD-pumped slab geometries at the forefront of the battlefield of laser development. The next steps are testing their performance and engineering such lasers for mobile platforms, i.e., aircraft, ships, and trucks. The USA Army announced it would move the 105 kW Northrop Grumman laser from the company's lab to the High Energy Laser System Test Facility at White Sands Missile Range (NM), where extensive testing takes place.

General Atomics is building a 150 kW laser for DARPA's High-Energy Liquid Laser Area Defense System (HELLADS), with a target weight less than 5 kg per kW of output in a laboratory demonstration version. The long-term goal is developing a laser weapon for tactical aircraft only.

Northrop Grumman is building the Maritime Laser Demonstration system, another LD-pumped solid-state laser is under demonstration at sea. Goals are very clear- to protect ships from threats such as small boats.

LD-pumping does not have a monopoly in tactical laser weaponry. The Navy is designing a 100 kW class free-electron laser weapon system for use on ships, in the hope that a free-electron laser could zap low-value targets at lower cost than expensive missiles. Planners envision eventual megawatt powers, which are very hard to imagine now on a board.

Fiber lasers may also have battlefield uses, although their limited powers so far seem best suited for detonating unexploded ordnance or land mines. But effective techniques for coherent beam combination could deliver more power to potential targets, expanding their uses. The most critical limitation for a fiber laser is the core diameter small size, which does not permit obtaining a high enough pulse energy for high-repetition-rate P-P operation.

Laser weapons have always been a tough challenge, and delivering a lethal dose of laser energy to the target has proved even harder than building the laser itself. The longer the propagation distance, the bigger the beam-delivery problem. So, it is not surprising to see military science shift its priorities to lower-power, shorter-range laser weapons. LD-pumped solid-state lasers can generate 100 kW outputs, powerful enough to block attacks by the rockets, artillery, mortars, and small boats favored by insurgents. Moreover, they can run from diesel generators that are already standard field equipment.

Major challenges remain. Optical damage is a critical issue, particularly in the inevitably dusty and dirty environment of a battlefield. Prototype solid-state lasers have used laboratory chillers to control thermal effects; practical versions will require rugged versions that can operate in the field. Other components also require adaptation for hard use, and costs must be reduced. Building super high power/energy lasers and delivering their beams to targets hundreds of kilometers away are problems that are even more formidable. LD-pumped alkali-vapor lasers are attractive because they promise very high efficiency, but flowing-gas versions might scale up to cover a few hundred of kilometers from only a single aperture. Generation of the same level of output from a slab-type solid-state laser would require a coherent combination of beams from many separate apertures; however, this technology is in its infancy. Thus, the future of that type of laser does not look very promising to overcome a megawatt level of an output power, which means that all above-mentioned technologies are limited and can be used for tactical laser weapon creation only.

New Approach to Development of a High-Power/Energy Laser

It is well known that CW mode of operation for high-energy lasers is very slow in terms of energy deposition into a target. Heat conductivity of a target, forced convection of surrounding media for a fast moving target, temperature dependable reflectivity of the material are not in favor of long time energy deposition. High-repetition-rate P-P lasing is the way to shrink the energy losses and to open a new set of "smart" interaction effects on a target. At the same time, it will help after transformation into a P-P regime to increase the range of effective destruction for the existing CW laser; in this case, an increase in the range by more than an order of magnitude is achievable! Conservation of the average power for the system during the process of transformation is possible, but peak power of laser pulses at the same time can be increased dramatically. Another very important feature of our approach is to avoid a well-known plasma screening effect on a target. Short enough pulses and effective matching of plasma decay time with repetition rate of laser pulses can significantly increase the rate of energy deposition into a target. Experimentally, all that had been demonstrated in our lab at the 100 kW level of average power. Peak power of the laser pulses was on the level of 10 MW. A similar approach could be valid in the case of the ABL COIL system, which after transformation of the resonator part only, can be much more effective and capable of locking the target at a distance of at least of an order of magnitude longer or to increase the peak power density on the target placed at the same distance as for CW mode for two-three orders of magnitude. According to our estimations, we should get after transformation: >500 km range of active interaction or much higher peak intensity >300 kW/cm2 at the distance about 50 km.

We are now looking toward a new generation of P-P lasers that could operate on vehicles at very high altitudes owing to advancements in laser technology, power generation and beam control at much longer distances. A new program should be established with an aim of employing a multi-ten-megawatt-class mono-module disk laser (MMDL) to clean space from debris and objects of natural origin. At the same time the general idea of the SDI in our days is going to be feasible as well. We should be focused on a new generation of laser systems with a much denser capability (at least an order of magnitude) or greater output power lasers in smaller packages (at least an order of magnitude as well) and operating at much safer conditions for the personnel involved. This will simplify future designs of many military systems and help make them much more effective. Existing high power laser systems based on electric-powered solid state multi-slab LD-pumped laser technology, however, does not look very promising for the particular task and cannot solve the problem of generating a high enough power/energy for the lasers to have operational benefit at significant ranges covering many hundreds of km.

Our idea of the MMDL works well. Modeling is done, amplified spontaneous emission (ASE) is effectively suppressed. Disk cooling technology by porous structures is not a problem at all and was demonstrated for high power optics long ago. Now is the time for the next step. It will be very expensive technological work. Hard to believe: Disk with $D = 50$ cm can extract about 10 MW of output power. Materials for the disk are ready! Scaling up technology is on the way. The MMDL can be combined with a telescope. Large apertures mean the absence of any problems with the coating damage and near optical surface breakdown effects. And so on... We do not see any negative physical or technical matters. "Zig-Zag" geometry or radiation phase-locking of many little disks has its limits. Our MMDL technology has no limits and can be applied to the level of an output power up to 40 MW. It was the goal of the SDI, but now this task has a more important target – space cleaning from manmade and natural debris. 52 years after the appearance of N.G. Basov`s idea of a disk laser, the problem of ASE suppression for the large disks had no effective solution [1]. Only now, we have found this innovative chance! However, this research will require substantial financing, taking into account the necessity of laser disk ceramics production with a size around 1 m. It is clear that cooperation with the US, Germany or Japan is the best way to have that work done during a reasonable time gap. The MMDL with an output of about 1 MW is the goal for the beginning!!! Scale up work to the level of 10 MW is our next step. Up to now, a very important part of the work has been done. We have made some preliminary estimations of the ceramics quality in comparison with the single crystal material. The data looks very similar, which is a nice characteristic of the background technology. The MMDL technology is scalable and can be used for different laser-based applications [2, 3]. Let us estimate the parameters of different MMDL scales for that special application, i.e., space cleaning: 100kW, 1 MW, 10 MW, 40 MW. The beam quality in our estimates was not very high, namely, 2 diffraction limits. Losses in the air (at a height of 2.5-3 km over the sea level) for the different interaction regimes and different track conditions were not taken into account.

At the bottom of the table one can see the data for low-repetition-rate pulsed DF (SIVD) laser, which was discussed in part III. The intensity level for the P-P regime is two-three orders of magnitude higher, which is dependable on time duration of pulses. The shorter the pulse duration and the lower repetition rate, the bigger the intensity of a P-P laser, as compared to the CW regime. For a DF laser this feature is the most critical at reasonable repetition rates of pulses.

No.	Average power/distance (Wavelength 1.06 μm) (2 dif.)	Diameter of telescope, m	Intensity kW/cm2 CW/P-P
1	Ty 100kW/40km	3	13 x 1000
2	1 MW/100km	3	20 x 1000
3	10 MW/500km	5	20 x 1000
4	40 MW/1000km	6	30 x 1000
5	15 kW/36 000 km SIVD (3.8 μm)	30	10

It is important to note that the level of intensities generated by lasers in the frame of "smart interaction" discussion is by many orders of magnitude higher than the "shut down" level of intensities for many types of sensors (IR and CCD) widely used in military equipment. A plasma pulsating discharge on a target due to the effective ablation process can be considered a very important factor of high-repetition-rate P-P laser technology. In conclusion, let us point it out that according to our estimates of the ratio (weight/power < 5kg/kW) and predictions of "Lockheed-Martin" for the disk laser technology, the ABL (2,5 MW COIL) can be replaced at the same board by 25 MW CW/P-P MMDL, which means a very high efficiency of packaging and energy extraction efficiency of the mono-module disk laser as itself.

REFERENCES

[1] Basov N.G., Bogdankevich O.V., Grasiuk A.Z., *IEEE J. of,* QE-2 (9), (1966).
[2] Apollonov V.V., *Natural science,* 4, 713, (2012)
[3] Apollonov V.V., *Natural science,* 4, 719, (2012)

Chapter 8

NEW APPLICATIONS OF HIGH-REPETITION-RATE PULSE-PERIODIC LASERS IN THE ARCTIC

INTRODUCTION

Works in the shelf zone of the Arctic Ocean will soon require the use of innovative, high-performance hardware equipped with high-power lasers. In this connection, the government of the Russian Federation intends to perform a wide range of activities associated with the active development of oil and gas deposits on the shelf in the Russian Arctic. These activities will involve the mastering of the technologies based on applications of high-power lasers that will reduce ice loads on engineering structures on the continental shelf and will provide an increased speed of ships and tankers following the Northern Sea Route. It is expected that high-power lasers can break the ice cover (up to several meters in height) around such engineering structures as oil platforms and icebreakers. The most effective in this case, is the joint application of high-power/energy lasers and icebreakers in the cracking of ice covering the ocean. For the first time, the feasibility of laser-assisted ice breaking in the interests of the icebreaker fleet was reported by Clark et al. at the conference of the National Bureau of Standards (Colorado, USA) [1]. This technology at present is being actively developed in Canada, USA, Norway, Sweden, South Korea, and Russia. It is expected that due to the positive test results that nobody questions, our country will start full-scale production of icebreakers equipped with high-power laser systems. It is known that experiments on the use of lasers in the Arctic have been conducted for many years and most new tried and proven technologies are already considered successful.

A laser installed on an icebreaker must comply with the requirements of the 'Russian Maritime Register of Shipping.' In particular, such a laser should be electromagnetically compatible with on-board equipment and comply with fire and radiation safety regulations. The weight of the laser system must be no more than 5 tons, and the time of continuous operation must amount to several tens of thousands of hours. The most dangerous hazards resulting from drifting ice can be easily overcome with the help of high-power/energy lasers. It is obvious today that lasers on offshore oil and gas platforms in the Far North will be in high demand and can greatly facilitate the safety of the personnel of such rigs.

WATER AND ITS STATES

Ice is one of the most enigmatic and simultaneously usual states of water. Water has a molecular weight equal of about 18.02, and may exist in liquid, gaseous, or solid states. Water is characterized by high boiling (+100°C) and melting (0°C) points, high values of heat of phase transitions [melting, vaporization (9.70 at 100°C, kcal/mol), sublimation]. Water has an anomalously high specific heat capacity, which is of great importance in nature: at night and during the transition from summer to winter water cools slowly, and in the afternoon or during the transition from winter to summer it slowly heats, thus moderating Earth's climate.

Water also has an unusual property of expanding when it is cooled down to its solid form; as a result, the density of ice is lower than that of water, which is not typical of other substances in the transition from the liquid to the solid state. This property of water creates a lot of problems for engineering structures that are in the aquatic environment. Among other anomalies of water one should mention the high surface tension and dielectric constant and a large thermal conductivity. The thermal conductivity of water is higher than other liquids and the thermal conductivity of ice is markedly higher than that of other non-metallic solids. The thermal conductivity of ice at 0°C is about four times higher than that water at the same temperature, i.e., ice conducts heat much faster than immobilized (not flowing) water in tissues. If at the same time we take into account that the thermal diffusivity of ice is much higher than that of water, it becomes clear why biological tissues freeze faster than become thawed. This property is fully applicable to the technologies of engineering construction under conditions associated with prolonged low temperatures.

The analysis of the well-known phase diagram of water states is important in the development of technological regimes. For example, if ice is heated at a pressure less than 0.610 kPa (4.58 mm Hg), it directly evaporates. This property is the basis for the development of the drying techniques for object and food preservation by freezing. One of the characteristics of water that distinguishes it from other substances is the reduction of the melting point of ice with increasing pressure, which is important when working at considerable depths of sea areas. Tables 1 and 2 show the basic thermal and mechanical properties of ice in a wide range of negative temperatures that are important for assessing the effectiveness of laser applications for its destruction by melting and sublimation (evaporation).

Table 1. Thermophysical properties of ice

t, ^{0}C	c_i, kJ/(kg K)	ρ_i, kg/m^3	λ_i, W/(mK)	a_i 10^6, m^2/s
0	2.12	916.4	2.23	1.2
–5	2.08	917.4	2.24	1.2
–10	2.04	918.2	2.25	1.2
–15	1.99	919.1	2.27	1.24
–20	1.96	919.96	2.29	1.3
–25	1.92	920.8	2.31	1.3
–30	1.88	920.8	2.32	1.34

Table 2 lists the physical data required for the analysis of various mechanisms of ice destruction. It should be noted that at low temperatures in the Arctic and Antarctic, purely mechanical destruction of ice is possible by using non-uniformly distributed laser radiation or electrical discharges, similar to the destruction of rocks due to thermomechanical stresses in mining or restoration of defective concrete structures in civil engineering.

Table 2. Physical properties of ice

Property	Values	Note
Heat capacity, cal/(g°C) Heat of melting, cal/g Heat of evaporation, cal/g	0.51 (0°C) 79.69 677	Heat capacity significantly decreases with decreasing temperature.
Thermal expansion coefficient, 1/°C	$9.1 \cdot 10^{-5}$ (0°C)	
Thermal conductivity, cal/(cm s·°C)	$4.99 \cdot 10^{-3}$	
Specific electric conductivity, $Ohm^{-1} \cdot cm^{-1}$	10^{-9} (0°C)	Activation energy 11 kcal/mol
Surface conductivity, Ohm^{-1}	10^{-10}(–11°C)	Activation energy 32 kcal/mol
Young's modulus, dyne/cm	10^{11}(–5°C)	Polycrystalline ice
Resistance, MN/m^2: to crushing to rapture to slicing	 2.5 1.11 0.57	 Polycrystalline ice Polycrystalline ice Polycrystalline ice
Average effective viscosity, P	10^{14}	Polycrystalline ice
Exponent of the power law of flow	3	
Activation energy under deformation and mechanical relaxation, kcal/mol	11.44–21.3	Increases linearly by 0.0361 kcal/(mol°C) from 0 to 273.16 K

LASER-ASSISTED ICE BREAKING

Currently, many technological processes make use of high-average-power lasers (100 W and higher), which operate in two modes, i.e., CW and pulse-periodic (P-P) regimes with a low pulse repetition rate (from a few to hundreds of Hz) and a pulse duration in the range of tens to hundreds of microseconds and milliseconds. In most cases, interaction of laser radiation with matter is accompanied by a purely thermal mechanism, because advantage is taken of the ability of a laser source to deliver a sufficient amount of energy to a small surface area of the processed material. Recently, fiber lasers have become very popular in various applications due to a number of very important advantages. However, in the case of fiber lasers, the thermal mechanism caused by the continuous mode of laser radiation and leading to the formation of a significant amount of the liquid phase, is not efficient. High-repetition-rate, high-average-power, P-P laser systems operating in the *Q*-switched regime, which provides a pulse duration in a periodic sequence from a few to hundreds of nanoseconds, allow one to implement a fundamentally different mechanism of radiation–matter interaction, i.e., sublimation of a substance to be treated, which provides a local release of energy not only in space but also in time [2–4]. It should be noted that in this case, the substance is evaporated explosively and locally, without an intermediate liquid phase. This mechanism can significantly extend the range of technological applications of laser sources. However, the high-repetition-rate regime of interaction has not yet found use in actual practice because of the high complexity of its implementation at high average powers required in many technological processes. In the early 2000s, we proposed for the first time and fabricated a high-repetition-rate, P-P laser having a power of 10 kW. The peak power of radiation generated by our laser was by several orders of magnitude greater than the average power of the laser source, i.e., the power that is typical of CW operation. The possibility of varying the pulse repetition rate and, hence, the value of the peak power has allowed us to control the interaction of radiation with matter over a wide range of parameters. Let us compare the two regimes (CW and P-P) from the point of view of their efficiency of ice cutting, i.e., possible thickness and cutting speed for a given power of the laser source.

It is known that in the case of CW radiation, the threshold power density of evaporation of ice is defined as:

$$q_{cw} = \frac{kT_{ev}}{Ar_0},$$

where A is the absorption coefficient at T_{ev} = 100°C, k is the thermal conductivity, and r_0 is the radius of the laser beam cross section in the focal plane. The resulting laser power is given by the expression:

$$P_{cw} = q_{cw}S,$$

where S is the area of the laser spot.

For a repetitively pulsed laser with the same parameters and the corresponding thermal regime, the threshold power density can be expressed as:

$$q_p = \frac{kT_{ev}}{2A}\sqrt{\frac{\pi}{a\tau}}$$

– a is the thermal diffusivity.

The pulsed power P_p and average power $\overline{P_p}$ can be expressed through the expression:

$$\overline{P_p} = P_p f\tau = W_p f,$$

where f is the pulse repetition rate, and W_p is the pulse energy.

Thus, on the basis of the above-said, we can generally conclude that

$$\frac{\overline{P_p}}{P_{cw}} = f\tau\frac{r_0}{\sqrt{a\tau}}$$

for a 1-kHz laser pulse frequency and a duration of 50 ns and higher. For the parameters mentioned, this ratio is equal to ~0.033.

Thus, for the repetition frequency of 1 kHz and pulse duration of 50 ns, i.e., the regime, which is sufficient for addressing most technological challenges, the ratio of the average repetitively pulsed power to the *cw* power turns out to be within a few hundredths. In other words, this means that cutting of ice having the same thickness and length requires the use of a laser whose power is 30 times less. In this case, we do not take into account the dynamics of the interaction of radiation with the products in the cutting zone, which in the case of *cw* operation is very power consuming.

Consider the process dynamics in terms of the explosive steam generation rate and minimization of the process of the liquid phase formation in the cutting zone in the case of high-repetition-rate, repetitively pulsed regime. The amount of the liquid phase (water in this case) is proportional to $\sqrt{a\tau}$, which corresponds to some zone heating by an individual laser pulse. The shorter the pulse duration at a fixed thermal diffusivity, the smaller the amount of water that prevents the penetration of laser radiation to the cutting zone. At the same time, the main fraction of the laser pulse energy transforms ice into steam, which, being explosively generated in the interaction region, carries away a small quantity of water that is simultaneously formed with the steam. The rate of the liquid phase removal from the interaction region (drilling and cutting) during the cutting of ice by the steam pressure is proportional to $1/\tau$. Both factors of the process obviously indicate the need to decrease the laser pulse duration.

For the high-repetition-rate, P-P regime, the pulse duration should be as short as possible, based on the requirements to the magnitude of the average power and the off-duty ratio of the pulse sequence generated by the laser.

Let us now compare the two generation (CW and P-P) regimes from the point of view of the requirements to the accuracy of the scanning system used for cutting. It is obvious that to maintain stable conditions for cutting ice (stable cutting quality on curvilinear segments, i.e., when turning the laser cutter on the surface of the ice field) it is necessary to provide a constant power flux density q. With respect to the average power of the high-repetition-rate, P-P laser, $\overline{P_p}$, this condition is expressed as: $\frac{\overline{P_p}}{V_{sc}} = \text{const}$. It follows that in terms of energy, the repetitively pulsed regime of laser operation is more suitable for cutting ice than the CW regime. This means that stable conditions imply constancy of the laser power $\overline{P_p}$ and the beam scanning velocity V_{sc}. This is particularly important for the cutting of ice along a complex trajectory. In this case, it is impossible to maintain the same velocity at linear sections and sharp turns of the trajectory. If V_{sc} is not constant and varies according to a certain law $V_{sc}(t)$, then in the case of a CW radiation source it is impossible to satisfy the condition $\frac{\overline{P_{cw}}}{V_{sc}} = \text{const}$ at any laser power.

It should also be noted that in the most general case, the physical processes proceeding during the high-repetition-rate, P-P cutting of materials are very similar to those during drilling in various engineering products. The cut is produced by a set of individual holes when the laser beam moves along the surface of the original object.

Data Analysis and Comparison with the Experiment

Clark et al. [1] presented data of ice cutting experiments for a CW CO_2 laser of 50 and 4500 W. The cited figure of merit for ice cutting, measured in $cm^2/(s \times kW)$, is the area of the cut, which differs in magnitude for the two power levels presented. These figures of merit were as follows: 8 $cm^2/(s \times kW)$ and 3 $cm^2/(s \times kW)$, respectively. Taking them as a basis, we assume the first number to be optimistic and the second – pessimistic. In the case of a CW CO_2 laser with an average power of 100 kW, we obtain for an 8-cm deep cut in ice the rate of about 1 m/s (optimistic estimate) and 38 cm/s (pessimistic estimate). In the case of a high-repetition-rate, P-P CO_2 laser with an average power of 100 kW, we obtain the following estimates for the same 8-cm deep cut in ice: the optimistic estimate amounts to 30 m/s and the pessimistic estimate – 11.4 m/s.

Based on the data obtained, it is important to pay more attention to the distribution of the laser power along an ice field in front of the icebreaker and to the geometry of the ship bow. Because laser radiation, cracking ice for the subsequent effective navigation through ice-covered waters, can be divided, using an optical system, into 2, 3 or more beams, the action of which must be accompanied by appropriate efforts of a specially designed bow of the ship or a set of metal structures, located on its bow. The above estimates clearly indicate that a 100-kW onboard laser system is sufficient to provide a large number of beams to crack ice and to meet many challenges that we will have to face during the development of the wealth of the Arctic Ocean.

Conclusion

Proceeding from the foregoing, one can conclude that for many processes requiring the processing and cutting of different materials (in particular, ice), the high-repetition-rate, P-P regime of laser operation seems much more efficient than the conventional CW regime. Fabrication of high-power/energy laser systems and their application to solve the stated problems will be an important national economic challenge that requires the full attention and support of the State.

References

[1] A.F. Clark, J.C. Moulder, R.P. Reed (1973). *Applied Optics*, 12 (6), 1103.

[2] V.V. Apollonov, V.V. Kiyko, V.I. Kislov, A.G. Suzdal'tsev (2003). *Quantum Electronics*, 33 (9), 753.

[3] V.V. Apollonov (2010). 'Impulsar': New Application for High Power High Repetition Rate Pulse-Periodic Lasers, Laser Pulse Phenomena and Applications, Dr. F. J. Duarte (Ed.), ISBN: 978-953-307-405-4, InTech, DOI: 10.5772/12832. Available from: http://www.intechopen.com/books/laser-pulse-phenomena-and-applications/impulsar-new-application-for-high-power-high-repetition-rate P-P lasers.

[4] V.V. Apollonov (2014). High Power P-P lasers (New York: NOVA Science Publishers Inc.).

Chapter 9

High Power/Energy High Repetition Rate Lasers for Space Cleaning

High Repetition Rate P-P Mode of Lasers Operation

For the lasers with a high average output power (GDL, HF/DF, COIL, Nd YAG), it is very common to use unstable resonator configurations having a large cross section of the active medium. In the resonators of this type, an externally injected low-power beam may exert a significant effect on the characteristics of output radiation.

One way to realize the radiation control regime is the self-injection regime of radiation, extracted from the resonator and returned back to the resonator as a part of radiation after changing its spatial-temporal characteristics [1, 2]. The transition to the transient lasing mode is effected through the modulation of the self-injecting beam. Earlier, a study was made of laser versions with radiation self-injection into the paraxial resonator region. However, analysis showed that the power of the beam injected into the paraxial beam region should be about the same value or comparable with the output laser power to efficiently control the resonator of a continuously pumped laser, unlike pure pulsed systems with regenerative amplification.

The self-injection of a part of output radiation through the resonator periphery is more efficient: on return to the paraxial resonator region, the injection power significantly rises due to the large number of passages to play the dominant part in the formation of output radiation.

In the case of a traditional resonator, the role of waves converging to the resonator axis was found to be insignificant, because their source is a narrow region with a small relative area at the edge of the output mirror; accordingly, the power of the control wave injected into the resonator is low. This wave has a large divergence, and only its small part (of the order of $1/N_f$, where $N_f >> 1$ is the Fresnel number) participates in lasing.

The effect of injection wave on the resonator characteristics can be enhanced by matching the beam phase with the resonator configuration and increasing the radiation power returned. In this case, the propagation direction and the wave front curvature of the injection beam should be so matched with the resonator configuration that the injection beam concentrates, after a relatively large number of passages through the resonator, near the optical resonator axis and transforms to a divergent wave that forms the output radiation. The injection beam energy should be high enough to exceed, after its arrival to the resonator axis, the saturation energy of the active medium.

P-P mode of operation was realized in two types of lasers theoretically and experimentally, in a gas-dynamic CO_2 and Nd YAG lasers [3]. CO_2–laser had the following parameters: the length of the active medium $L_a = 1.2$ m, the unsaturated gain coefficient $g_0 = 0.6$ m^{-1}, the time it takes the active medium to transit the resonator $T = 0.92*10^{-4}$ s, the relaxation time $T_p = 2.76*10^{-4}$ s, the total go-round resonator time $T_f = 4.2*10^{-9}$ s, the luminescence lifetime $T_l = 5$ s, the resonator magnification factor $M = 1.45$, the diameter of output laser aperture a=0.08 m. Nd YAG – laser was above of 1 kW level, with two heads geometry.

The CO_2–laser resonator is made up of two spherical mirrors with rectangular apertures, which provided a geometrical amplification factor of 1.45. The active medium travels across the optical resonator axis. In what follows, all theoretical and experimental data are given for a laser with the above parameters.

A part of the output laser radiation was diverted by an inclined metallic mirror to the injection beam formation system consisting of two spherical mirrors with conjugate focal planes. In the vicinity of the focal plane there formed the waist of the branched part of the laser beam, and a modulator was placed near the waist. The modulator location was so selected that the laser beam completely filled the aperture of the modulator. The maximal modulation frequency in our experiments has reached 50 kHz.

A mirror 4 focused the radiation onto the calorimeter. The duration of an individual pulse was about 100-150 ns. We emphasize that the recorded pulse duration was limited by the measuring path bandwidth equal, as noted above, to 50 MHz. The amplitudes of individual pulses exceeded the average value of output power by factor –10. The average output power was measured with a calorimeter cooled by running water. It is noteworthy that the average output power in the pulse-periodic mode was equal to the output power in the CW laser-operating mode. Good agreement between the experimental and theoretical data for frequencies ranging up to 25–50 kHz testifies to the adequacy of the proposed model and the possibility of employing this method at higher frequencies to convert a CW laser to the operating mode similar to the Q-switching mode.

HF/DF-laser and COIL are waiting for the experimental efforts to be applied. Theoretically P-P modes of regenerative amplification for high power lasers have been investigated and modeled by computer. The output parameters are very much dependent on parameters of media, way of pumping and resonator geometry. The summary of the radiation temporal structure is presented:

COIL P-P mode starts at frequencies >20 kHz.
Depth of modulation -100% at frequencies >100 kHz.
Pulse duration <250 ns. Ratio P peak./P aver. = 100-1000.

HF/DF P-P mode starts at frequencies >100 kHz.
Depth of modulation -100% at frequencies >250 kHz.
Pulse duration <100 ns. Ratio P peak./P aver. = 100-10000.

Nd YAG P-P mode starts at frequencies >4 kHz.
Depth of modulation -100% at frequencies >40 kHz.
Pulse duration <250 ns. Ratio P peak./P aver. = 100-1000.

CO P-P mode starts at frequencies >10 kHz.

Depth of modulation -100% at frequencies >100 kHz.

Pulse duration <250 ns. Ratio P peak./P aver. = 100-1000.

NEW APPLICATION FOR HIGH REPETITION RATE P-P LASERS

In recent years, increasing attention has been given to studying the possibility of lasers used for cleaning the space from elements of space debris (ESD), which have been collected during more than four decades of operation of space, which, in some cases, create a serious threat to space vehicles (SV). By expert estimations of space, by 1996 there was about 3.5 million not traced ESD in the size less than 1 cm, more than 100 thousand splinters in the size in a diameter from 1 to 10 cm, the size nearby 8000 ESD exceed 10 cm [4-6]. Large ESD with a diameter more than 10 cm are found by modern watch facilities and are bought in special catalogues. The most effective method of protection from such ESD is maneuvering of SV. By estimations of experts, splinters less than 1 cm in diameter do not represent special danger to existing SVs because of the presence of passive constructional protection, though it makes SV considerably heavier. The most unpleasant diameter of splinters is 1 … 10 cm, when the necessary degree of passive protection does not manage to be carried out because of its unacceptably large weight, and avoiding collision by maneuvering SV is impossible, as on the radar screen such splinters are not visible.

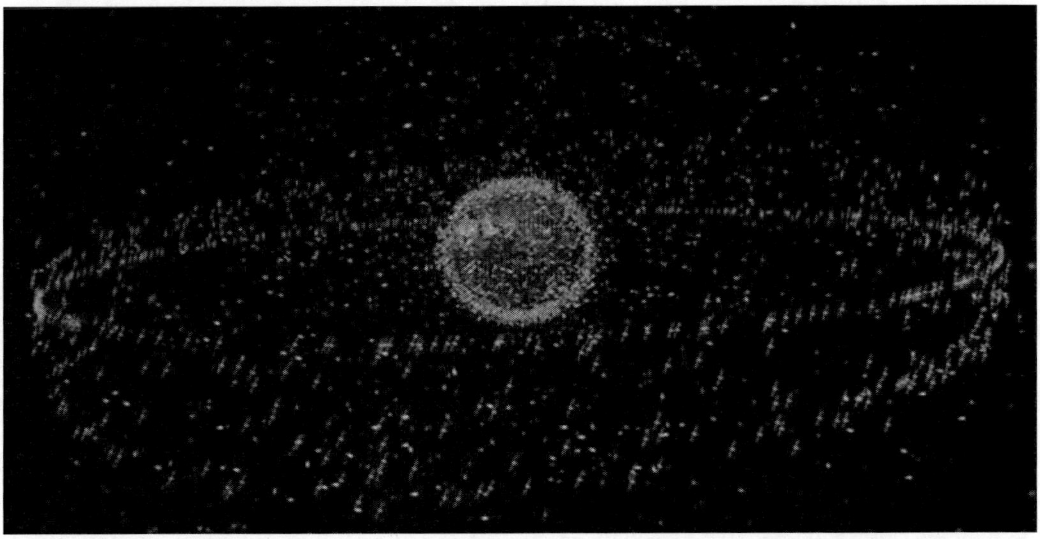

Figure 1. Space debris around globe.

In low orbits under the influence of atmosphere, quickly enough there is their self-clearing as time of life ESD in orbits with height about 200 km averages about one week. In higher orbits in height, their self-cleaning can occupy of 600 km 25-30 years, and at heights about 1000 km – 2 thousand years [5]. The estimations executed in work [ ], have shown that the probability of collision SV in diameter of 10 m within one year of its operation makes 0.45 10^{-2} for ESD with a size 2 … 4 cm and 0.4 for ESD with a size <0.4 cm, and frequency of collisions with the catalogued objects (≥10 cm) is at level of one collision for 30 years. Every year the number ESD regularly increases. From here the reality of threat of collision with ESD for all periods of operation, the SV is clear.

Thus withdrawal ESD from an orbit protected SV is an actual problem. For this purpose, it is necessary to reduce the speed of its movement. As it will be shown further, it is possible to reach by pulse irradiation ESD and reception on its surface of the plasma creating an impulse of return. Such impulse, arising in a mode of laser ablation of ESD material, should reduce height of orbit ESD so that it has flown by SV or, finally, would enter into dense layers of atmosphere and has burnt down.

In a number of earlier works, it was offered to use the Nd YAG ground-based laser installations for space clearing, but such laser has the essential lacks connected with necessity of passage of 1 mkm radiation of the big capacity through atmosphere that can lead to loss of optical quality of a bunch of radiation and occurrence of nonlinear effects They have small mobility, therefore the number of ESD, which are possible to influence by their radiation, will be limited. At influence of laser on ESD from the Earth surface, the return impulse will be directed upwards and the apogee of orbit ESD will be increased, but the perigee is going to be decreased and stopped by the dense atmosphere. And the most important thing – requirements to power the land-based laser – should be increased in comparison with the space laser, as the distance from a terrestrial surface to ESD is much more. For these reasons, the most expedient arrangement of Nd YAG 100 kW laser installation directly in space is recommended. Thus, it is desirable that in such installation, power consumption should be minimal. This condition satisfies Nd YAG with LD pumping, capable of working independently in P-P mode of operation with very small expenses of power for the system service control. But for the case of the ground-based laser, we have suggested the DF-laser, which radiation propagation through the atmosphere is much more effective and the output power of existing systems (>1.5MW) and technologies are more advanced.

In the paper [5], the most possible variants of rapprochement ESD, flying, as a rule, on elliptic orbits, with various SV, moving on circular orbits at heights 200-700 km, have been analyzed. Two variants, when SV moves on a circular orbit at height of 400 km have appeared the worst, and ESD fly on elliptic orbits with height of apogee – 2000 km and – 4000 km. In this case in a perigee there are areas where planes of orbits SV and ESD coincide, and speed of their rapprochement is maximum, and in this area vectors of speed SV and ESD lie along the same direction, i.e., by influence of the laser radiation it is impossible to give ESD a lateral component of speed, as in more opportunity of an inclination of planes of orbits SV and ESD under the relation to each other.

The maximum speeds of rapprochement calculated for these two variants have made accordingly – 395 m/s and – 2463 m/s. For circular orbits with height 200, 400 and 700 km settlement of rapprochement ESD speeds, flying on circular orbits, with SV do not exceed 343 m/s, therefore these variants can be neglected.

Let's consider process of rapprochement ESD, which is catching up to SV, after influence on ESD of the laser radiation. Before the laser influence, the force of an attraction of the Earth and centrifugal force are equal:

$$\frac{m\upsilon_0^2}{R+H} = \gamma\frac{mM}{(R+H)^2},$$

where υ_0 – speed of movement ESD on a trajectory before influence of a laser impulse, R – radius of the Earth, H – height ESD over the Earth, M – weight of the Earth, γ – a gravitational constant, m – weight ESD after such influence on ESD this balance will be broken; then the reduction of ESD speed $\Delta\upsilon$ will force the normal acceleration in the direction to the center of the Earth:

$$a_{\text{H}} = -\frac{\gamma M}{(R+H)^2} + \frac{(\upsilon_0 - \Delta\upsilon)^2}{R+H},$$

where $\Delta\upsilon$ – change of ESD speed after laser pulse influence (typical value $\Delta\upsilon$ makes ~ 200 km/s [7]). After simple transformations from (1), we will receive

$$a_{\text{H}} = \frac{-2\cdot\upsilon_0\cdot\Delta\upsilon + \Delta\upsilon^2}{R+H}.$$

Through the time – t the radius-vector of ESD orbit will be changed:

$$\Delta H = \frac{a_{\text{H}} t^2}{2} = \frac{\Delta\upsilon\cdot(\Delta\upsilon - 2\upsilon_0)}{R+H} t^2$$

By knowing the initial distance – L from SV to ESD and a tangential component of rapprochements speed of ESD to SV after influence of a laser pulse:

$$t = \frac{\ell}{\upsilon_{\text{т}}} = \frac{\ell}{\upsilon - \Delta\upsilon}$$

Then for change of size of a radius-vector of ESD orbit, we will receive the following expression:

$$\Delta H = \frac{\Delta\upsilon\cdot(\Delta\upsilon - 2\upsilon_0)}{R+H}\frac{\ell^2}{(\upsilon - \Delta\upsilon)^2}$$

From here it is possible to find the distance between ESD and SV when it is necessary to start the influence on ESD by laser:

$$\ell = (\upsilon - \Delta\upsilon)\sqrt{\frac{\Delta H\cdot(R+H)}{\Delta\upsilon\cdot(\Delta\upsilon - 2\upsilon_0)}}$$

Proceeding with the SV dimensions, we will be set by size ΔH = 30 m. Then for the first variant of rapprochement ESD to SV at V_{rapp} = 395 m/s, $\Delta\upsilon$ = 200 m/s, V_0 = 8 km/s, H = 400 km, R ≈6300 km the distance between will make 4.1 km. This way will be passed in time ~

20 s. Then for metal ESD at typical values $C_m^{опт}$ = 4 din-s/J and S/m = 0.15 cm2/g we will receive $\Delta\upsilon$ = 6 cm/s, and the necessary number of pulses for the value $\Delta\upsilon$ = 200 m/s will make 3300 pulses at frequency of high repetition rate Nd YAG laser – 3000 Hz necessary time of influence 1,1s, which is much less than time of rapprochement to SV found before (20 s). It shows that with the same laser it is possible to reject ESD from SV with rapprochements having much greater speed.

For more exact calculations at the big speeds of rapprochement, it is necessary to consider dynamics of change of values $\Delta\upsilon$ and current distance between ESD and SV after influence of each laser pulse of P-P irradiation of ESD.

For the second variant with very great speed of rapprochement V_{rapp} = 2463 m/s at $\Delta\upsilon$ = 200 m/s and much bigger distance – 20 km the rejection is possible as well. However, maintenance of $\Delta\upsilon$ at the distance – 20 km will meet some changes of parameters due to the bigger size of the focal point on such a distance.

The problem of ESD withdrawal from SV orbit that ESD has flown by SV has been considered above. Another problem is important also – to create such impulse of return to achieve decrease ESD to an orbit in height of 200 km when at the expense of the further braking in atmosphere of particles of ESD will be burned out, and the space will be cleared from ESD. In other words, SV with laser installation will carry out onboard a role of "cleaner" of the most used orbits. That particle ESD has decreased to 200 km over the Earth surface, its speed needs to be reduced by certain value $\Delta\upsilon_{clean}$, which will allow it to pass from a circular orbit on elliptic, which exact value can be calculated as follows:

$$\Delta\upsilon_{clean} = V_{apogee} - V_{start},$$

where V_{apogee} – ESD speed in apogee of a transitive elliptic orbit, V_{start} – speed ESD in an initial circular orbit. Speed in apogee is:

$$V_{apogee} = \sqrt{\frac{2 \cdot \gamma \cdot M \cdot R_{start}}{r_{200} \cdot (r_{200} + R_{start})}},$$

where r_{200} – radius of a circular orbit in height of 200 km, R_{start} – radius of an initial orbit. ESD speed in an initial circular orbit is defined as:

$$V_{start} = \sqrt{\frac{\gamma \cdot M}{R_{start}}}.$$

On the basis given by A. V. Avdeev et al. calculations, the graphic dependence of demanded reduction of speed ESD in apogee of an elliptic orbit from height of an initial circular orbit has been constructed. The similar dependence has been resulted in the work [5] without explanations. It is clear that ESD, being in the orbit with height ~900 km, will decrease it to the height of 200 km, if the speed is reduced by 200 m/s.

Change of ESD speed $\Delta \upsilon$ after the influence of laser radiation pulse with energy density E [J/cm^2] on ESD is defined from the following expression:

$$\Delta \upsilon = Cm\, E\, S/m;$$

where S – the interaction area, m – weight of ESD, Cm [din s/J] – proportionality factor between $\Delta \upsilon$ and E, depending on the ESD type. Characteristics of the most widespread of them are presented in Table 1 [5]. Such ESD are formed as a result of SV explosions, or their collisions with ESD. Spheroids of Na and K are formed after destruction of reactors, splinters of phenol-carbon plastics and fragments of "plastics-aluminum" are the fragments of thermal protection; splinters of aluminum-based materials can appear after explosion of tanks and covers of SV; steel bolts – fragments of connecting blocks armature.

Table 1. Cm(opt) and S/m for different ESD

	Type of ESD				
	Na (K)	"C"- based material	Organics-based material	"Al"- based material	"Fe"- based material
Angle, (degree)	65	87	99	30	82
Apogee, (km) Perigee, (km)	930 870	1190 610	1020 725	800 520	1500 820
S/m, (cm^2/g) size, (cm)	1.75 1.0	0.7 1×5	2.5 0.05×30	0.37 1×5	0.15 1×10
Reflectivity	0.4	0.02	0.05/0.7	0.05/0.7	0.5
$C_b^{opt}, \left(\frac{Din\ s}{j}\right)$	(6 ± 2)	(7.5 ± 2)	(5.5 ± 2)	(4 ± 1.5)	(4 ± 1.5)

Figure 2. Space based laser for fast re-entering of the ESD to the dense layers of atmosphere.

High power high repetition rate P-P laser should generate a temporally, and spectrally effective pulse designed for high transmission through the atmosphere, as well as for efficient ablative coupling with the target.

The space based Nd YAG laser with output power less than 100kW that we propose is the best tool for fast re-entering of the ESD to the dense layers of atmosphere (Figure 2).

The DF ground-based laser system that we have proposed is capable of rapid engagement of targets whose orbits cross over the site, with potential to kill on a single pass. Very little target mass is ablated per pulse so the potential to create additional hazardous orbiting debris is minimal.

The laser system would need to be coupled with a target pointing and tracking telescope with guide-star-like wave-front correction capability.

Table 2 presents the LEO/MEO ESD removal data for Nd YAG laser. ESD have a size 1 - 10 cm and fly below 300 km altitude. Cm = 4 dyn-s/J in average for polymer and "Al"-based materials response. Typical S/m data for ESD: NaK-1,75; Al-0,37;Fe-0,15 are taken from the Table 1. For I= 3.0 J/cm2, S/g = 0,15cm2/g, we need N =7000 laser pulses for ESD re-entry. Nd YAG - laser operating at 3000 Hz can re-enter a small object from the gap 1-10cm in 2.3 s. Such a level of average output power (360kW) for CW/P-P Nd YAG lasers has not been demonstrated up to now. To get such effective results for clearing, we do not need the laser but only a 30 m in diameter telescope to deliver the laser pulses to a target at 300 km range or more with 10 ns time duration:

Table 2. LEO/MEO ESD removal data for Nd YAG laser

λ	τ	D_b	W	f	<P>	d_s	Z	I
1.06 μm	10 ns	30 m	60 J	3000 Hz	360 kW (0,5)	5,2 cm 2Dif	300 km	3,0 J/cm^2

Table 3. LEO/MEO ESD removal data for DF laser

λ	τ	D_b	W	f	<P>	d_s	Z	I
3.8 μm	10 ns	30 m	150 J	10 KHz	1,5 MW	18 cm 2Dif	300 km	0,6 J/cm^2

Table 3 presents the LEO/MEO ESD removal data for DF laser. ESD have the same size 1 - 10 cm and fly below 300km altitude. Cm = 4 dyn-s/J in average for the same materials: polymer and aluminum. With I = 0.6 J/cm2, t = 10ns, S/g = 0,15cm2/g, we need N = 35000 pulses for ESD re-entry. Ground based 1,5MW DF - laser operating at 10 kHz can re-enter any small object from the same size gap in 3,5 s. This operation requires a 30 m diameter telescope to deliver 2 J/cm2 (Cm = 0,2 Cm opt) to a target at 300km range with a 10 ns pulse at 3.8 mkm: Here it is important to note that with one minute delay for retargeting all objects of this height and below can be re-entered during - 0,5 year only. It should be noted as well here that the level of output power for CW regime had been demonstrated and the technology is mature enough. The realization of P-P mode of operation for this type of laser is the question of time. Motivation is completely available. New tasks for high repetition rate high power lasers generated during the last few years are very important [7-9] and should definitely be resolved during the upcoming years.

Conclusion

1. In the paper the questions of SV protection and of clearing orbits from dangerous ESD with diameter from 1 to 10 cm by means of high power/energy high repetition rate P-P Nd YAG with average power 100kW and DF-lasers with average power about 1,5 MW are considered;

2. Possibility of applying above-mentioned installations not only for dangerous ESD withdrawal from SV orbit, but also for planned clearing of the most maintained orbits from these ESD when the given installations will carry out a role of "cleaner" of these orbits is analyzed. For this purpose, under the influence of radiation it is necessary to translate ESD from a circular orbit to elliptic, which perigee is in the dense atmosphere beds where ESD should be burned down. As a result of the decision of a ballistic problem, dependence of necessary reduction of speed ESD from height of their orbit over the Earth is received. It is found that for orbits with heights up to ~300 km, the demanded influence can be provided by 1,5MW DF-laser installation with a telescope with diameter -30m and duration of pulses of about ten nanoseconds;

3. It is shown that for the worst variant in case of influence on metal ESD with the greatest speed of their rapprochement with SV ~2,5 km/s angular divergence of radiation of space based 100kW Nd YAG laser should be not worse than two diffraction limits with use of a telescope with diameter of main mirror $D = 1$ m is admissible.

References

[1] Apollonov V. V., Vagin Yu. S., “Conducting channel creation in a non-conductive media”, Patent № 2009118874, 2009.

[2] Apollonov V. V., Alcock A. J., Baldis H. A.,*Opt.Letters,*5,333,1980.

[3] Apollonov V.V., Kijko V.V, Kislov V.I., High-frequency pulse-periodic regime of powerful radiation in the wide-aperture lasers, *Quantum Electron,* 2003. T. 33. Number 9. p. 753-757.

[4] *Phipps C., Michaelis M.M.* LISK-BROOM: Clearing near-Earth space debris in 4 years using a 20 kW, 530-nm repetitively pulsed laser. Conf. "Physics of Nuclear Induced Plasmas and Problems of Nuclear Pumped Lasers". September 26-30, 1994. - Arzamas, Russia.

[5] Campbell I.W. Project ORION: orbital debris removal using ground-based sensors and lasers//NASA Technical Memorandum 108522. 1996.

[6] *Phipps C.R., Luke J.R., Funk D.J., Moore D.S., Glownia J., Lippert T.* Measurements of Laser Impulse Coupling at 130fs//Proc. SPIE. 2004 V. 5448. P. 1201-1208.

[7] Apollonov V.V. Reports of the Academy of Sciences (DAN), V. 351, Number 3. p. 339-340, 1996.

[8] Apollonov V.V, Kijko V.V, Kislov V.I, Tischenko V.N Pulse-periodic lasers for Lightcraft/Impulsar application, *Proc SPIE,* V.5777. P.1011-1018,2005.

[9] Apollonov V.V, Grachev, G.N, Gulidov A.I. Interaction of an optical pulsed discharge with a gas: conditions stable generation and merging of shock waves, *Quantum electronics*.T. 34, Number 10, p. 941, 2004.

Chapter 10

LASER TECHNOLOGY AND WEAPONS

INTRODUCTION

The experience of designing strategic laser systems in the previous years and a strong belief in the attainability of the goal – the development of high-power/energy laser weapons – help to significantly accelerate the pace of work in the field of new technologies. World press has recently published articles where it was suggested that the anti-missile airborne laser project suffers cost overruns and delays and may fall victim to budget cuts. Just after that mass media immediately reported that the Pentagon refuses to develop combat lasers. As a result, there has appeared an epic set of myths about combat lasers. The task of this paper is to consider the most ridiculous statements and to reject them by sufficient analysis. An additional goal is to focus the interests of the laser physics community to the mono-module disk laser geometry, which is now possible for implementation.

The Wall Street Journal has recently published an article in which it was suggested that the anti-missile airborne laser (ABL) project suffers cost overruns and delays and may fall victim to budget cuts. Mass media immediately and happily reported that the Pentagon refuses to develop combat lasers. However, this euphoria seems very misleading, if not tougher.

Today, chemical lasers are being widely replaced by solid-state diode-pumped lasers. It is these lasers that the Pentagon counts on, because they are more compact, simpler and cheaper to use than chemical lasers. Besides, they are reliable, easily compatible (without any transformation of the output voltage) with nuclear and solar energy, allow for further scaling of output parameters and the efficiency of their operations is significantly higher. Northrop Grumman Corporation has already presented a 105-kW solid-state laser and intends to significantly increase its output power. According to data from U.S. laboratories, work is on the way to develop a prototype of a 500-kW laser. Subsequently, 'hyperboloids' will be truck-mounted [high energy laser technology demonstrator (HEL TD) program], ship-mounted [maritime laser demonstrator (MLD) project], and airborne [high energy liquid laser air defense system (HELLADS); a laser for F-35, B-1, CH-47 aircrafts). Another direction is largely supported by Raytheon, which stakes on fiber systems. A 50-kW laser weapon system (LaWS) will be integrated with the Phalanx Close-In Weapons System (CIWS) against anti-ship missiles and its land based version of Centurion C-RAM. In addition, the officials in the

U.S. have recently reported on the progress of works on a combat free electron laser. At the same time, we should not forget about "Alpha" system (a laser with an output power of 4.5 MW), lying on the ground and waiting for the decision to be launched.

Figure 1. Anti-missile airborne laser (USA).

Bringing solid-state lasers to megawatt powers takes time and considerable resources. However, the experience of designing strategic laser systems in previous years and a strong belief in the attainability of the goal – the development of high-power laser weapons – helps to significantly accelerate the pace of work in the field of new technologies. It should be noted, however, that tactical laser systems with lower power outputs have already been tested in the U.S. and now find applications in the army. All this testifies that the Pentagon experts clearly do not think about cancelling the promising laser programs. We are talking here about an effective system of disinformation. Last year's report 'Technology Horizons' by the Pentagon refers to the global changes in the 'rules of the game' after the proliferation of 'high energy weapons', which will turn the traditional symbols of military power into outdated stuff like cannonballs and cavalry... And while the U.S. develops laser programs, many other countries show 'laser apathy.' Incompetent bloggers and pseudoscientific workers, who had something to do with the laser program thirty years ago, speculate that 'lasers are a bluff.' As a result, there has appeared an epic set of myths about combat lasers. Consider the most ridiculous of them.

2. Ten Myths

2.1. Myth 1. "Combat Lasers Have Been Developed for Four Decades with No Progress in Sight"

Let us quote one of the Russian papers: "In the 1970s the Americans took a 150-ton Boeing-707 and 'stuck' a laser to it, which successfully burned small rockets. In the 2000s, they took a 350-ton Boeing-747 and 'stuck' a heavier and more powerful laser, which successfully burned larger size rockets. In 20 years they will buy in Ukraine a decommissioned "Antonov AN-225 Mriya" (640 tons), and here it is, the Death Star [2]. Yes, probably it will be able to shoot down a "Scud", and even a "Taepodong". However, only at the site, and once, no more."

Under the '150-ton Boeing-707,' to which a laser was 'stuck,' is apparently meant a 137-ton Boeing KC-135 stratotanker based on the" 707 "of the first), modified in 1973 into NKC-135ALL [airborne laser laboratory (ALL)]. In 1983, a laser mounted on the aircraft shot down several Sidewinder air-to-air missiles at a distance of 5 km and some other 'little' things. What has changed since then? According to the above quotation – only the size of the aircraft.

Figure 2. Air based military laser.

But what about the reality? Even the so-called 'megawatt' CW lasers of the 1980s did not emit light at megawatt power levels, they consumed more power. The 2.2-megawatt Miracle laser system, which later in the combat version was called a tactical high-power/energy laser (THEL) ('MIRACLE' with the 'SEALITE' guidance system) did not possess a supernatural power [3]. What to say about the earlier and five times weaker ALL. Has there been any progress since then? An ABL has a power of 1.1 MW and it is not the power to be consumed; this is the power in the beam. Thus, a 'more powerful' (50 times) laser was 'stuck' to a 350-ton Boeing... However, it should be understood that the actual capabilities of the laser rely not on the power as such, but on the concentration of the radiation, i.e., the ability of the 'guns' to emit not only a powerful, but also a narrow beam. The ALL had the level of the radiation concentration equal to 10^{13} J/(sr·s). An ABL has the level of about 10^{18} J/(sr·s) – that is, 10 thousand times higher. These achievements are made up not only of the straight-line growth in the power. The last 30 years have seen a period of extremely rapid development of adaptive optics to compensate for the effects of atmospheric turbulence and laser path on the transmitted beam. In addition, the lasers of the same class are radically reduced in size. The first version of a THEL weighed 180 tons and could hardly be installed into six trailers. The laser used a hydrogen–fluoride mixture, which is very unfriendly to the environment. The second generation of advanced tactical lasers (ATLs) relies on oxygen–iodine mixtures [so called chemical oxygen–iodine lasers (COILs)] and is more compact. Finally, a new solid-state laser of Northrop Grumman weighs 1.5 tons, including the cooling system. In the future its weight is expected to be reduced to 750 kg. As a result, the land based version of the system consists of a heavy expanded mobility tactical truck (HEMTT A3), command post on a high mobility multipurpose wheeled vehicle and a towed single-axle trailer with the

AN/MPQ-64 radar. At the same time, the U.S. works hard to convert the CW regime of the lasers in the P-P one, which will dramatically increase their range of actions.

Figure 3. Mobile high power/energy laser system.

Talks about the fact that "combat lasers have been developed for forty years and so they are hopeless," only show ignorance in technical matters. Breakthrough technologies are first tested for a few decades before the entry into a phase of maturity. Thus, aircrafts at the time of the first flight had almost 60 years of history – the first flying models were built in 1840 and the attempts to build full-size airplanes date back to 1868. It is, in fact, a classical scheme of development of any technology using new physical principles: First, a long 'incubation period' with no apparent practical results, and only then a 'great leap forward.'

2.2. Myth 2. "Lasers Cannot Be Used for a Long Time, Usually They Work Several Seconds"

It is not so! In fact, chemical and solid-state combat lasers ensure continuous wave operation for minutes and tens of minutes. The next step in the development of high-power laser systems, of course, will be the implementation of a variable temporal structure of radiation in order to raise the peak radiation power for the ablation mechanism and to eliminate the effect of the screening of the target by plasma [4].

2.3. Myth 3. "Energy of Laser Weapons Is Negligible Compared to the Fire Arms"

For comparison, the power of a 76.2-mm F-22 divisional gun (1936) is 150 MW. This power is 150 times higher than that of the ABL! Besides, we ignore here the energy of the explosive in the shell, which is the same. Ponder upon this simple fact: A small ancient gun of the Second World for the price of scrap metal is hundreds of times more powerful than the ultramodern 'combat' laser weighing tens of tons and valued at over five billion USD. Only one shot from the ABL costs millions of dollars. And this shot is comparable in energy with the burst of a machine-gun fire.

Such a comparison of the power achieved in 0.01 is with the power of CW light, and the 'proof' of inferiority of 'long-play' weapons (by using this comparison) contradicts even the school physics course. Let us try to compare everything in the correct way, i.e., by counting the energy sent to the target.

The muzzle energy of a 12.7-mm heavy machine gun is 15 – 17.5 kJ at a rate of 80 – 100 shots per minute. In other words, a 100-kW laser can be 'replaced' by three and a half heavy machine guns (6000 kJ/min vs. 1750). But let us return to the gun. The muzzle energy of F-22 is 1.35 MJ, while the power of the ABL is 1.1 MW, i.e., 1.1 MJ every second. Thus, the laser shoots 48 'shells' per minute. By converting the MW power to the TNT equivalent, we will obtain 240 grams of explosives per second and 14.4 kg per minute, which corresponds to the content of 18 high-explosive shells from the same gun. However, the actual 'value' of the laser is higher. The matter is that even with accurate firing of firearms, most of the 'energy' goes not to the enemy, but dissipates. The reason for that includes a dozen factors (wind, fluctuations in humidity, air pressure and temperature, the Coriolis force, etc.), making the bullet or shell spread inevitable. A steady flux of photons flies to where it was sent, excluding the huge amount of unnecessary loss.

2.4. Myth 4. "Efficiency of Lasers Is a Few Percent"

In fact, the efficiency of combat lasers is about 20.6%, and this is not the limit. According to the RELI program, the efficiency is planned to be raised up to 25%. Fiber lasers, which are adapted by Raytheon for military applications, now have an efficiency of about 30%. The efficiency of firearms is 20%–40%. At the same time, the efficiency of (smaller but steadily growing in power) solid-state diode-pumped systems is more than 50%, and soon it will approach its physical limit of about 85%.

2.5. Myth 5. "The Laser Beam Has a Huge Diffraction Divergence"

Here we deal with a physically insuperable law of diffraction, which states that the laser beam always diverges with an angle proportional to the ratio of the wavelength to the diameter of the beam. If we take a specific combat IR laser with a wavelength of 2 μm (this wavelength is typical of combat THELs, etc.) and a beam diameter of 1 cm, we obtain a divergence angle of 0.2 mrad (this is a small angular divergence; for example, standard laser pointers/rangefinders have a divergence angle of 5 mrad and higher). However, because of

0.2-mrad divergence the beam spot diameter will increase from 1 cm to 3 cm at a distance of 100 meters. That is, only 100 meters away the beam density will decrease (proportionally to the area of interaction) by 7 times. At a distance of a kilometer, the density of the beam falls by 300 times.

Actually, a combat laser emitting a beam with the initial diameter of 1 cm is a fruit of an unhealthy imagination, not burdened with some knowledge in this area. In fact, when using the focusing optics the diffraction divergence is about λ/D, where λ is the wavelength, and D is the diameter of the mirror and also the initial beam diameter, tapering due to focusing as the beam approaches the target; a large beam diameter (meters in this case) provides a low diffraction divergence. In the case of the ABL the wavelength is 1.315 μm, and the diameter of the mirror is 1.5 m; by dividing one by the other, we obtain the divergence of about 10^{-6} rad. In other words, at distance of only one kilometer the laser beam from the Boeing diverges by (oops!) 1 mm.

At a distance of 200 km, the diffraction divergence determined spot will be about 20 cm. The actual ABL beam divergence exceeds the diffraction limit by only 1.2 times.

2.6. Myth 6. "One Can Easily Be Protected from Laser Weapons Using, for Example, an Aluminum Mirror"

Yet, another malapropism: Indeed, some metals may have near-100% reflection coefficients. First, however, these coefficients are not equal 100% equal. Thus, at a wavelength of 1 μm the reflection coefficient drops to 75% for most structural metals. A real missile after its launch will be also significantly contaminated with combustion products. Meanwhile, modern 'hyperboloids' emit light in the vicinity of 1 μm (the ABL has a wavelength of 1.315 μm). In this case, 25% of hundreds of kilowatts will be sufficient even in the cw regime to heat up and melt down the thin skin layer of the missile. Thus, reflectivity will decrease because the absorption of laser radiation increases rapidly with increasing temperature, and abruptly jumps after the start of melting. In the pulse-periodic regime the situation is even more favorable.

Figure 4. Different constructions of laser mirrors.

Besides, there arises also a ‘childish’ question: If the laser beam can be focused and directed by a mirror, why can’t the same mirror protect us from the laser beam? In lasers, highly reflective multilayer dielectric mirrors are usually used – but in a very narrow range and only under strictly defined angles [5]. In addition, they are cooled, which is impossible to do with the entire surface of the target. In other words, simple, effective and affordable protection from high-power/energy lasers does not exist.

Figure 5. Water cooled through micro-channels mirror with protective case.

2.7. Myth 7. “The Problem of Overheating of Lasers Cannot Be Solved”

Four megawatts of heat, which can heat an aircraft to red-hot and burn it to the ground, are generated per each megawatt of laser power. The cooling system with the gas flow rate of 1800 m/s (de Laval nozzle) is unable to release the heat generated from the fuselage.

In reality, the ‘disposal’ of the amount of heat in the megawatt units is quite trivial. Has anyone seen a red-hot diesel locomotive? Meanwhile, a decent diesel engine with a capacity of 2 MW releases more than 1 MW of heat into in the oil and cooling system. Far less easy is the problem of heat release from the limited volume of the laser weapon. In the case of a chemical ABL, the heated reaction products are simply blown out of the cavity by the well-known de Laval nozzle, and then liquid ammonia is used for cooling. Fairly cumbersome is the system with cryogenic components; however, it really is able to ‘recycle’ a very impressive amount of heat. Tactical solid-state lasers, which need more than 400 kW of heat to be released, do well without cryogenic ‘refrigerators.’ Thus, the HELLADS makes use of a unique cooling technique; the circulation of a liquid releases excess heat outside the ‘laser cannon.’ Remarkable also is the General Atomics’ advanced thermal energy storage device, capable of cooling directed energy weapons. Heat is stored in the 35 kilogram module by melting a wax-type phase change material. As a result, the HELLADS provide for the interception of missiles within two minutes at the specified cw power followed by a thirty-second break.

2.8. Myth 8. "High-Power and Compact Energy Sources for Combat Lasers Do Not Exist"

This is partially true: It is not yet possible to mount a 100-kW solid-state laser on anything less than a truck because of the need to have at hand a 500-kW generator and suitable capacitors. In fact, this is a real problem, which has nothing to do with fantasy. In practice, the hybrid version of the HEMTT (HEMTT A3), even in the basic version has a 350-kW generator, which can provide up to 200 kW. When the engine power is increased up to 505 hp, HEMTT A3 can produce 400 kW of continuous exportable power. A nice addition is the 1.5-MJ capacitor bank. In other words, where the bloggers fancy an electric power plant, in fact, there is a high-tech truck. However, the issue of energy in the space can be solved by other, more efficient ways.

For example, nuclear power sources, and solar energy with its unlimited possibilities, are relatively well developed.

Figure 6. Mobile laser system.

2.9. Myth 9. "Every Shot of a Laser Is Worth Millions of Dollars"

In fact, one shot of the ABL costs 10 thousand USD, whereas the domestic 16 million rubles are the propaganda and exaggeration. This is comparable with the cost of a Fagot anti-tank guided-missile system. More complex anti-tank guided-missiles cost tens of thousands of dollars; for example, a Maverick air-to-ground missile for hitting targets ranging from a distance of a few thousand feet to 13 nautical miles at medium altitude costs 154 thousand USD, a Patriot missile – 3.8 million USD. The cost of a shot from tactical lasers is less than from the ABL. Even in a hydrogen fluoride THEL it was 2-3 thousand USD; in this case, the laser used was not hydrogen but deuterium, which is quite expensive.

Figure7. Sketch of space based laser system.

2.10. Myth 10. "All the Problems that Can Be Solved with Laser Weapons, Are Easier and Cheaper to Solve by Traditional Means"

This conclusion has already been proven to be ineffective; see, for example, Israel's attempts to defend against missile attacks by Hamas missiles (the Iron Dome system). One interception costs 30-40 thousand USD. The cost of a rocket for a Grad multiple rocket launcher system is about 1 thousand USD, the cost of a Qassams rocket does not exceed 200 USD. Thus, the interception will be 40-200 times more expensive than the rocket launched by the enemy. In this connection Tariq abu Nazar, the Hamas spokesman, once said "if every rocket launch of our missile men costs tens of thousands of dollars for Israelis, we will assume that the goal is achieved." As a result, some newspaper people accuse not developers of laser systems but those who are responsible for termination of the Israeli-American program. The Centurion missile defense system has not found wide application because of the short-range and great ammunition consumption.

3. Legends

Of course, this is not a complete list of the legends about lasers. Most of them are built on the same principle: either an outrageous lie or painstakingly making a mountain out of a molehill. In fact, lasers on the battlefield are real and the army, which possesses such weapons, will receive an impressive advantage. Thus, the air force, able to actively defend against anti-aircraft and air-to-air missiles, will be much less vulnerable to air defenses. Thus, the development of laser technology is critically important, and not only for well-developed countries. Combat lasers are an obvious asymmetrical response to the superiority of the precision weapons. The 'ideology' of the latter in a very rude manner is to ensure that instead of a dozen of shells, a guided munition (though much more expensive) is used to precisely hit a specific target, and to minimize collateral damage. However, such a scheme is particularly vulnerable to laser defense systems, which make no difference between an archaic shell for two hundred dollars or an expensive high-tech device or military equipment. The number of targets is not so great, and their cost is ten times higher than that of the most expensive laser 'shot.'

4. Disk Laser

It is already evident that the world has entered a new round of technological race. It does not depend on our wishes; it is the willing of time. Most developed countries, based on their technological advantages, are spending billions of dollars to develop high-tech next-generation lasers [6]. According to Japanese media, up to now, the U.S. has invested more than 100 billion USD in the development of semiconductor laser pumping of solid-state lasers.

The mono-module disk laser concept is one of the most effective designs for diode-pumped solid-state lasers, which allows the realization of lasers with super-high output power, having very good efficiency and also excellent beam quality. Since the first demonstration of the principle in 1966 [7], the output power of mono-module disk has been increased to the level of few kW in continuous wave (CW) mode of operation. "Zig-Zag" disk laser geometry does not look like a prospect for further growth of output parameters. The scaling laws for mono-module disk laser design show that the limits for CW mode of operation is far beyond 100 kW for output power and the energy can be higher than 100 J in pulsed mode of operation. Due to the efficient porous cooling technology and possibility of amplified spontaneous emission (ASE) suppression, the operation of the mono-module disk laser geometry is possible in CW and P-P modes at extremely high output power.

This is due to the indisputable advantages of disk geometry in terms of the minimal thermal lens in the active media and the high radiation resistance of the disk in the P-P regime because of the large area of the optical surface to couple out the radiation. So, the necessity to find a solution to the problem of the ASE suppression along the diameter of the disk was the major problem (matter of patent). In our case, the size of the disk at a multi-megawatt level of the average power output should be at least 50 cm, i.e., at least hundred times bigger than the size of the disk that is used today in the existing systems. Radiation from such a laser, obtained during generation in the active medium of a single disk, does not require additional phase-locking. At the same time, such a laser in a mono-module geometry will be very well combined with a large-diameter telescope for ensuring high peak power density of the laser pulse on space debris. It is known that the disk geometry of a laser was proposed 51 years ago; however, to this day, the solution to the problem of the ASE suppression with increasing transverse dimensions of the active medium in the mono-module disk geometry is found! Thus, the prospects of new versions of the mono-module disk laser creation for new class of cutting edge problems are open!

Conclusion

The need to accelerate the development of high technology is the major topic of many speeches by political leaders. It is important to note the opinions of Western experts, who say that today laser technology is one of the most effective ways to gain technological superiority in the world. And today, laser technology makes it possible:

- to provide a new level of development of industry, science and technology for your own country, the revival of scientific and technological strengths of the country on the basis of modern high-techs;
- to make your country a leader in the field of technical, scientific and technological progress;
- to revive a large number of enterprises in many sectors of industry, which are well known for their developments in previous years;
- to strengthen your country's leading position in space and to ensure strategic priorities in the modern world;
- to derive benefit from the sale of a wide range of laser programs, which is today comparable to the benefit from the sale of traditional products and resources.

References

[1] SPIE Events Europe (2012), Nothrop`s R@D, INTERNET.
[2] P.V. Zarubin, QE 32(12), (2002).
[3] *Journal of Aerospace and Defense Industry News,* Dec.1997.
[4] V.V. Apollonov at. all, QE 33(9), (2003).
[5] V.V. Apollonov, *Laser Physics* 23, (2013).
[6] V.V. Putin, *Message to the Federal Assembly of RF,* Dec.12, (2013).
[7] V.V. Apollonov, *Natural Science* 5(5), (2013).

Chapter 11

POWER OF ATMOSPHERE

There are many appearances in the literature of reliable observations of studying so-called "jets" and "sprites" – the discharges in the gigantic natural capacitor "Ionosphere-Earth". The volume of such a discharge is approximately 5-10 thousand cubic kilometers and it usually appears above the surface of ocean. There are also cases of above-mentioned discharges on the ground. The value of the energy transferred to the earth can comprise several tera-joule. Events are accompanied by the emission of the waves of ultra-low frequency. Their study has significant interest from many points of view. The essence of the observed irregular phenomena consists of the electromechanical conversion of the energy excesses of natural electricity into mechanical and thermal energy of cyclones, typhoons and other natural cataclysms. The ionosphere can retain only the specific quantity of energy. Otherwise, it discards the surpluses of electricity through the atmosphere or transforms them into the energy of storms, in that number and inside the Earth. By using the part of the natural electricity for useful purposes, it is possible to govern the weather of the planet. Causing the artificial breakdown of the ionosphere, it could be possible to arrange the discharge of the controlled aqueous sediments at the necessary points of the globe. It could also be possible to attempt to regulate the climate of the planet and to decrease the amplitudes of magnetic storms, earthquakes and hurricanes.

Lightning is the electrical spark discharge, which is manifested, usually, by the bright flash of light and by its accompanying thunder. The electrical nature of lightning was opened in studies of the American physicist B. Franklin, according to idea, which was carried out, regarding the experience on the extraction of electricity from the thunderstorm cloud. Famous Russian scientists: M. Lomonosov and G. Rikhman also dealt with this problem Aside from Earth, lightning had been located on the surface of other planets: Venus, Jupiter, Saturn and Uranus. The average length of lightning is 2-5 km, and some discharges stretch in the atmosphere up to the distance of 150 km. Let us look at the process of the appearance of lightning. Most frequently, the lightning appears in the rain clouds, due to that they are called thunderstorms. Sometimes lightning can be formed in the layered rain clouds, and also with volcanic eruptions, tornados and dust storms. Usually linear lightning is observed, which relates to the so-called electrode-less discharges, since they begin and end in the accumulations of the charged particles. This determines some not so clearly explained properties, which distinguish lightning from the discharges between the electrodes. Thus, lightning does not occur shorter than several hundred meters; they appear in the electric fields considerably weaker than the field with the inter-electrode discharges; the collection of the charges, transferred by lightning, occurs for a thousandths of a second from a huge number of small and well isolated from each other particles, located in the volume of several km^3. The process of the development of lightning in the thunderstorm clouds has been most studied, in this case the lightning can pass to the clouds themselves – intra-cloud lightning, and they can strike into the earth – ground-based lightning. For the appearance of lightning it is necessary for the relatively small, but critical, volume of cloud to be formed in the electric field with tension, sufficient for the beginning of the electrical discharge ~ of 1 MV/m, and in the substantial part of the cloud, there would be a field with the average tension, sufficient for maintaining the discharge ~ of 0.1-0.2 MV/m. In the lightning the electrical energy of cloud is converted into the thermal and the light. The process of the development of ground-based lightning consists of several stages. At the first stage in the zone, where the electric field reaches critical value, impact ionization begins, created by at first free electrons, always existing in a small quantity in air, which under the action of electric field acquire significant speeds in the direction to the earth and, colliding with the molecules, which compose air, they ionize them. Thus, they appear the electron avalanches, which pass in the thread of electrical discharges – streamers, which are the well conducting channels, which, merging, give beginning to the bright thermo-ionized channel with the high conductivity – to stepped leader of lightning. The motion of the leader to the earth's surface occurs by steps into several ten meters with a speed of ~50,000,000 meters in a second, after which its motion stops by several ten microseconds, and the glow strongly weakens; then in the subsequent stage leader again moves by several ten meters. A bright glow covers all passed steps; then followed again by stoppage and weakening of glow. These processes are repeated during the motion of leader to the earth's surface with an average speed of 200,000 meters per second. The field strength at this end is strengthened in proportion to the advance of the leader to the earth, also under its action from the objects protruding on the earth's surface when the reciprocal streamer, which is connected with the leader, is ejected. This special feature of lightning is used for creating the lightning rod. In the final stage on ionized by leader channel follow reverse (from bottom to top), or main thing, discharge of lightning, which is characterized by currents from ten to hundreds of thousands of amperes, by the brightness, which noticeably exceeds the brightness of leader, and with the high speed of advance, which at first reaches to ~ 100,000 kilometers per second, while in the end of that decreasing to ~ 10,000 kilometers per second. The temperature of channel with the main discharge can exceed 25.000°C. The length of the channel of lightning can be from 1 to 10 km, diameter – several centimeters. After pulse advancing of ionization current of channel and its glow are weakened. In the final stage, the current of lightning can last the hundredth and even tenths if logged, reaching hundred and thousands of amperes. Such lightning is called protracted, and they most frequently cause fires. Main discharge discharges the part of the cloud frequently. The charges, located at high altitudes, can begin a new arrow-shaped leader, who moves continuously with a speed of thousands of kilometers in second. The brightness of its glow is close to the brightness of stepped leader. When the arrow-shaped leader reaches the earth's surface, it follows the second main attack, similar to the first. Usually lightning includes several repeated discharges, but their number can also reach several tens. The duration of repeated lightning can exceed 1 s. The displacement of the channel of repeated lightning by wind creates the so-called strip lightning – luminous strip. With the entry of lightning directly into the soil, the formation of the unique mineral of fulgurite is possible, which is, in essence, the sintered quartz sand. Intra-cloud lightning usually includes only leader stages; their length varies from 1 to 150 km. The portion of intra-cloud lightning grows in proportion to displacement to the equator, changing from 0,5 in the temperate latitudes to 0,9 in the equatorial strip. The passage of lightning is accompanied by changes in the electrical and magnetic pour on and by the radio emission, by the so-called atmospheric disturbances. The kill probability by the lightning of ground-based objects grows in proportion to an increase in its height and with an increase in the electrical conductivity of soil on the surface or at a certain depth. The action of lightning conductor is based on these factors. If there is an electric field in the cloud, sufficient for maintaining the discharge, but insufficient for its appearance, the role of the initiator of lightning can be carried out by long wire cables, products of the combustion of the engine of aircraft or rocket. Specifically, this sometimes "provokes" lightning in the layered rain and cumulus congestus clouds.

For the illustration of enormous power engineering of the lightning and the significant frequency of their repetition on the terrestrial globe, let us review an article from the British press about "Marakaibo`s lighthouse" (Venezuela).

There are more than three million discharges of lightning i.e., more than a thousand for each second that occurs each day in the world. The greatest frequency of discharges occurs in the tropical and subtropical zones of the terrestrial globe, the enumeration of these are record in the following places: Ale hook in Columbia (270 days in the year with lightning); Tororo in Uganda (274 days); Bogor in Java in Indonesia (283 days). Completely extraordinary light storms occur in Venezuela in the mouth of Katatumbo river, which falls into Marakaibo Lake. This spectacle can be seen at night for more than ten hours, one-hundred-eighty days of the year. Light shows occur exclusively in complete silence because of the intercloud discharges of nature at the height up to 10km altitude. The sky is illuminated with bright flashes, approximately 300 times an hour, which comprise more than one million electrical discharges per year, with current strengths from 100 to 400 thousand amperes each.

Figure 1. Experience Venezuela's Catatumbo Lightning that Strikes 1 Million Times Per Year. The atmospheric phenomenon lights up the sky.

These discharges are visible from the ocean for many tens of kilometers. Over many centuries, seafarers used them as a natural lighthouse, which was called “The Lighthouse of Marakaibo”. Mixing cold wind from the Andes with the hot and humid air, which is raised from the swampy earth's surface in the environments Marakaibo is the major factor, which creates this unique phenomenon. Air rising from the surface of swamps in essence consists of methane, which appears due to rotting of plants. Each flash of lightning could illuminate all the lamps of South America. Now all depends on humanity, which must find the method of the domestication of this enormous energy.

LIGHTNING IN THE NATURAL CAPACITY “EARTH – CLOUD”

Why does it thunder and then sparkle lightning? The clouds accumulate electric charges, which leads to the breakdown of air, which is accompanied by the flash of light and by the formation of shock acoustic waves. In other words, thunder and lightning are the manifestation of atmospheric electrical existence. The fundamental connection of thunderstorm phenomena with electricity, for two and one-half centuries, was reliably confirmed by numerous studies of geophysicists, meteorologists, and specialists on lightning protection. However, a qualitative jump in the improvement of the methods of observations in the last two decades led to the unexpected discoveries in the study of thunderstorm clouds, lightning, and the orbital scale of atmospheric electrical chains. This interest in the problems of atmospheric electricity will make it necessary to look differently at the mechanisms of shaping the electric field and its role in the dynamics of atmosphere and ionosphere of the Earth. The heart of the atmospheric electrical machine – the thunderstorm cloud, is more precise, the totality of simultaneously “working” thousands of thunderstorms, distributed in the lower part of the atmosphere to the troposphere. Thunderstorm clouds do not live for too long – from an hour to several hours. But the change to some thunderstorms come others, forming in the troposphere on the proximity. Contemporary satellite measurements, and also ground-based systems of the registration of lightning, give researchers the sufficiently reliable maps of the allocation of frequencies of lightning flashes over the Earth's surface. Flash rate above the surface of the ocean on the average, is lower than above the continents in the tropics. The reason for this asymmetry is the intensive convection in the continental regions, where the land is effectively heated by solar radiation. The quick ascent of thoroughly heated air, saturated by moisture, contributes to the formation of powerful convective clouds of vertical development, in the upper part of which the temperature is lower than ~40°C. The result is the formation of particles of ice, soft hail, hail, whose interaction against the background of the rapid ascending current leads to the separation of charges. Cloud height in case of a large water surface, on average lower than above the continents, and the processes of electrification is less effective above the ocean. Lately another factor discussed is the difference in concentrations of aerosols above the ocean and the continents. Since aerosols serve as the condensation nuclei, necessary for forming particles in the supercooled air, their abundance above the land increases the probability of the strong electrization of cloud. The quantitative analysis of this factor requires detailed experiments. Approximately 78% of all lightning are recorded between 30° S. and 30° N. The maximum value of the average density of the number of flashes per unit of the earth's surface exceeding 80 $1/km^2$ is observed in

Africa (Ruanda). The entire pond of the Kongo River, with an area of about 3 mln. km^2, regularly demonstrates the greatest lightning activity, although other equatorial centers abound with thunderstorms. It seemed that the subtropics and middle latitudes have the most significant global thunderstorm activity. Some of them, for example in North Argentina and Paraguay, lead in the rate of lightning flashes. The study of the regional special features of statistics requires further improvement of the satellite and ground environment of mapping thunderstorms. These studies are intensively conducted in the connection with the realization of lightning activity as an important factor of the climate of planet, which sensitively reacts to changes in the temperature, humidity, radiation conditions, and the composition of the atmosphere. The progress in studies of atmospheric electricity is connected, first of all, with the study of the mechanisms of generation and dissipation of electrical energy in the atmosphere. The main problem here – the so-called electrical dynamo: the generation of quasi-stationary electric fields and space charges in the weakly moving conductive medium. Already the simplest formulation of the problem of dynamo is manifested in its important special feature – very wide intervals of time-spatial scales. So that on the scale of several kilometers would arise the electric field, compared to the value of the breakdown of ~30 kV/cm for dry air under the standard conditions, necessary that the random traffic of charges with the collisions of cloud solid or liquid particles would lead to the matched effect of the addition of micro-currents into the macroscopic current of the completely high value (several amperes), sufficient for the rapid (tens of seconds) process of the separation of charges. It is evident from the experience that in the ripe thunderstorm cells the lightning discharges occur with enviable regularity – each 15-20 s, i.e., the mechanism of charging acting in the cloud is very effective, although the average electric charge density rarely exceeds several nCl/m^3. As shown, the measurements of the electric field on the earth's surface and also inside the cloud medium (on the balloons, the aircraft and the rockets), in the typical thunderstorm cloud "main" negative charge – on the average several ten coulomb – occupies the interval of heights, which corresponds to temperatures from –10 to –25°C. "Basic" positive charge is also composed of several ten coulomb, but it is located above a basic negative; therefore the large part of the lightning discharges the cloud-earth returns to the earth negative charge. The lower part of the cloud frequently reveals a smaller value positive charge of ~10 Cl. to explain the 3-layer structure of field and charge in the thunderstorm cloud described above. We will examine the set of the mechanisms of the separation of charges. They depend on such factors as temperature, phase composition of medium, spectrum of the sizes of cloud particles. The dependence of the value of the charge from the electric field transferred for one collision is very important. From this parameter it is accepted to subdivide all mechanisms into the induction and the non-inductive. For the first class of mechanisms, the charge q depends on the value and direction of the external electric field and is connected with the polarization of the interacting particles. Non-inductive of the charge exchanges between the being encountered particles in the explicit form on the field strength does not depend. In spite of the abundance of different microphysical mechanisms of electrization, now many authors consider the main thing to be non-inductive charge exchanges colliding with small sizes of one to ten micrometers of ice crystals and particles of soft hail with the dimensions of the order of several millimeters. Experiments established the presence of the characteristic value of a temperature, at which the sign of charge q reverses, points of reverse, which usually lay between –15 and –20°C. Specifically, this special feature made this mechanism so popular, since the typical temperature profile in the cloud explains the 3-layer structure of the distribution of the charge density. Recent experiments showed, however that many thunderstorm clouds possess an even more complex structure of space charge. Mesoscale is especially interesting, which means the horizontal scales from ten to hundreds of kilometers, the convective systems, which serve as the important source of thunderstorm activity. Their characteristic feature is the presence of the united electrical structure, which includes the region of intensive convection and sometimes extends to several hundred kilometers in the stratified region. In the region of stratification, the ascending currents are sufficiently weak, but the electric field has a steady multilayer structure. The zero isotherm near here is sufficiently narrow, with thickness into several hundred meters, and the stable layers of space charge, which are mainly critical for the high lightning activity of mesoscale convective systems. The mechanism and laws governing the formation of the layer of positive charge in the environment of the zero isotherm remains debatable. Other examples of the work of electrical dynamo serve electrization and discharge phenomena with the dust and snowy storms, with the volcanic eruptions, nuclear explosions in the atmosphere, in the technological processes, connected with mixing of aerosol flows, for example in the flour-grinding and cement industry. The presence of powerful convective and turbulent flows in the thunderstorm clouds, and also the above cases of electrization, indicate that interaction of the large charged particles with the sizes of 0.1-10 mm with the particles of micrometer size in the moving conducting gas medium plays an important role for the electrical dynamo. The poly-phase and multi-flow nature of medium brings in the final sum to the effective transformation of mechanical energy into the electrical. The problem of the initiation of lightning discharge remains sharp and unites the entire complex of questions.

Let us briefly pause at two of them. First, as measurements on the balloons, the rockets and the aircraft showed that the maximum tension of the electric field in the thunderstorm clouds does not usually exceed 2 kV/cm, which is substantially lower than the threshold of the breakdown of dry air at the heights in question, the composing value order 10 kV/cm. In the thunderstorm cloud, however, the discharge is developed in the humid air, which contains particles of different sizes. It is natural to assume that the threshold field is reached in the compact spatial domain, for example, on a scale less than 10 m or generally on the microscale, because of field strengthening on the particles, and in the sufficiently short period, so that the usual sensors of field do not ensure the required three-dimensional – time resolution with the registration. The development of the avalanche of the fast, so-called running away electrons with energy on the order of 1 MeV can serve as the alternative explanation to initiation of sparks in the sufficiently weak field. Such electrons can be accelerated in the cloud field because of a drop in brake power with an increase in the energy of particle. The threshold of the development of avalanche is almost an order lower than the usual threshold of the breakdown of dry air; therefore when the seeding particles of the high energy are present, which are supplied by cosmic rays, the development of avalanche in the cloud can give birth to the localized region of high conductivity, capable of initiating spark. In the second place, the classical theory of gas discharge does not make it possible to explain the rapid passage of cloud medium into the conducting state at the preliminary stage of lightning. Recently the new scenario of this stage had been proposed and investigated, it is connected by reaching by the cloud of the regime of the self-organized criticality. In the model of electrical cells with the significant dimension of ~1-30 m and with that random growth in space and time by the potential separate small-scale breakdown between the pair of cells, capable of causing a “chain reaction" of intra-cloud micro-discharges, the stochastic process “of the metallization” of intra-cloud medium is developed. This model describes well the experimentally observed special features of the preliminary stage of the lightning discharge, in that the number dynamics of micro-discharges are characteristic of radio emission. The study of a thunderstorm phenomena includes experiments on the artificial initiation of lightning – trigger lightning. In order to cause the discharge in the atmosphere under the thunderstorm cloud, the rocket, which leaves after itself in the regime of a deficiency in the oxidizer, a significant quantity of soot conducting electricity is used. As it is proved to be, initiation occurs with the sufficiently high tension of field on the earth's surface under the cloud – not less than 5-10 kV/m. The for the first time a classical diagram of the initiation of discharges with the wire pulling after the rocket was realized in 1960 from onboard a research vessel. From those times, thousands of successful experiments have been carried out; physics of the ascending and arrow-shaped leaders, the return shock, mechanism of the sudden strengthening of the luminosity of lightning channel because of the sharp current amplification was investigated, in detail, because of them. Today there are new possibilities for controlling lightning discharges, in that number with the aid of lasers. Thus, powerful lasers are capable of organizing the extensive plasma cylinders in air, which could initiate and direct the lightning discharges similar to metallic wire in the trigger lightning and conducting channels from the products of the combustion of special geodetic rockets. In 1925, Ch. Wilson indicated the possibility of accelerating the electrons to the relativistic energies in the field of thunderstorm clouds. This hypothesis was confirmed during our days: on the generation of relativistic particles and quanta of high-frequency emission in the atmosphere, which correlates with the thunderstorm activity, a whole series of experiments testifies. Thus, measurements on the balloons revealed an anomalous increase in the intensity of X-ray and gamma-radiation of more than to two or three orders in the course of approximately one minute with the maximum spectrum in the region 50-60 keV. Ground observations also revealed the intensive flashes of X-radiation with the energy of quanta into several hundred keV, whose appearance coincided with molding of the leader of lightning flash and the arrow-shaped leader of trigger lightning. Finally, in the latter several years were published the results of observations from the satellites of the splashes of the gamma-quanta MeV-foot level, X-ray and ultraviolet radiation of atmospheric origin. To explain these phenomena, a number of new works have drawn the phenomena mentioned above, for the atmospheric electrodynamics, and precisely: the generation of the electrons running away and the breakdown of these electrons. Energetic quanta can arise from of the electronic radiation of the fast electrons, which interact with the molecules of air. Models describe different situations, including the development of the avalanche of fast electrons when the extensive air shower of cosmic rays is present, and breakdown on the electrons running away in the strong field of the stepped leader of lightning. One should note that an increase in the number of fast electrons in the field of thunderstorm cloud with the presence of an extensive air shower is accompanied by the generation of a large quantity of second particles, and this leads to the generation of the current pulses and radio emission. If the energy of the primary particle is sufficiently great (10^{17}-10^{19}eV), the short (several microseconds) pulse of radio emission can have enormous energy (to 1 MeV), which explains the appearance of the so-called narrow bipolar pulses, observed sometimes during ground-based and satellite measurements and correlating with the thunderstorm activity. The formation of such intensive current pulses is of interest both for understanding the mechanism of the generation of lightning and for studying the cosmic rays of super high energies. It is interesting that the presence of the cellular structure of the electric field in the thunderstorm cloud with the amplitude, which exceeds the threshold "of running away", proves to be essential for the process of accelerating the electrons to the relativistic energies. The randomly oriented electrical cells, together with the acceleration, sharply increase the lifetime of relativistic electrons in the cloud because of the diffusion nature of their trajectories. This makes it possible to explain the significant duration of the splashes of X-ray and gamma-radiation and the nature of their interrelation with the lightning flashes. The role of cosmic rays for the atmospheric electricity must explain experiments in a study of their correlation with the studied thunderstorm phenomena.

LIGHTNING IN THE NATURAL CAPACITOR "CLOUD – IONOSPHERE"

The powerful pulse to the study of atmospheric electricity at the very end of the 20th century gave observations of the discharge phenomena in the middle atmosphere – Sprites, which correlate with the thunderstorm activity, each day where the region of glow stretches to 85-90 km above the earth, and flash duration comprises from ones to tens of milliseconds, elves, who stretch to the heights – 70-90 km with the duration less than 100 s and jet – discharges, which start from the upper part of the clouds and which are extended at times to the mesospheric heights with a speed of about 100 km/s. Registration of high-altitude discharges and diagnostics of their characteristics are complicated because of the short time of life and sporadic nature of the lightning discharges. Partly therefore, until now the physical models of Sprites, Elves and Jets remain the object of intensive discussions. Perhaps, the greatest information to the present time is accumulated about Sprites. Jet and its physical examination is at the most interesting stage of work for the future.

The optical flash of Sprite in the mesosphere appears through several milliseconds after discharge cloud (+) – the earth, sometimes at the removal of several ten kilometers along horizontal from the channel of lightning. Discharge ignites at a sufficiently high altitude, since the threshold of the breakdown of air falls with the height exponentially, whereas the amplitude of the disturbances of the electric field, which appear at the moment of the powerful lightning flashes of capacitor "cloud - the earth", decreases with height much slower – according to the power law, and at the heights approximately 75 km in the atmosphere exceeds the threshold of breakdown. In the last decade, the study of the high-altitude discharges in the atmosphere became extensive, intensively developing in the direction of geophysical electrodynamics. And although the stage of the accumulation of data, that characterizes the morphology of these phenomena, is by no means completed, it is already possible to switch over to a study of the thinner special features of structure and dynamics of the high-altitude discharges and their role in the orbital electrical chain and the balance of low components of the atmosphere. Experiments and theoretical calculations show that these discharges must be considered as the part of the dynamic process, which includes the formation of the structure of field and charge in the cloud and lightning discharges to the earth. Different approaches to the simulation have been widely discussed, including the attraction of the mechanism of the running away electrons. Details were developed in the model of the generation of electric field in the middle atmosphere, which considers the special features of the charge distribution and its dynamics in orbital scale, the convective systems, which serve as the basic sources of positive discharges to the earth. The physical model had been developed, which makes it possible to describe fine structure and dynamics of Sprite. Sprite is represented as the network of micro-discharges – streamers – and as the self-sustaining process in the external field. Together with metallizing process of the thunderstorm cloud mentioned above, we have one additional example of self-organizing, when the dynamics of high-altitude discharge is caused by reaching the threshold of the so-called directed flow, which characterizes the formation of the branched conducting channels, which overlap the entire length of Sprite. Immediate prospects in the study of the high-altitude discharges in the atmosphere are connected with conducting of the coordinated ground-based and space experiments, and also with the development of new methods of diagnostics of the parameters of high-altitude discharges, in that number with the use of possibilities of radio-wave and laser diagnostics. Electrical field of the atmosphere is very variable. The tension of the vertical component of field, which is usually much more than horizontal, reaches several kV/m with the saggings, the ground-winds and the thunderstorm cloudiness. Therefore, the concept of conditions introduces prettier weather, which corresponds to wind speed not more than 6 m/s into the absence of any kind of sediments, hoarfrost, fog, lower cloudiness. But even under these conditions near the earth's surface there is an electric field of tension approximately 150 V/m, along which in the weakly conducting air flows the electric current with the density several pa/m^2. This field changes in time and the space, the fluctuations relative to average value can comprise from one to ten percent. The measurements of electric field, current and conductivity under the conditions allow better weather where they serve as the powerful means of the study of the electrical state of the atmosphere. However, it is possible, having only learned to divide global i.e., planetary scale, orbital - with the scale of the order of the height of neutral atmosphere 100 km and the local disturbances of the electrical parameters. The latter, in their turn, are directly connected both with the changes in the ionic-aerosol composition and with the dynamics of medium. In the studies, it was established that the boundary layer of the atmosphere is characterized by the presence of the aeroelectrical structures, which are manifested in the pulsations of the electric field with the short-term power-law spectra with periods from one to several hundred seconds. The electrodynamics of fog is of special interest. Full-scale observations showed that under the conditions of fog in the frequency band 10^{-2} – 1 Hz also realize the power-law spectra of the pulsations of field, but the intensity of pulsations increases more than by the order in the comparison with the conditions grow prettier weather, which is explained by the diffusion charging of drops in the lowest layer of the atmosphere. The results of structural – temporary analysis made it possible to isolate two forms of the electrical state of fog, the first of which is characterized by the formation of intensive aeroelectrical structures, and the second – by chaotic structural – temporary variations in the field and current.

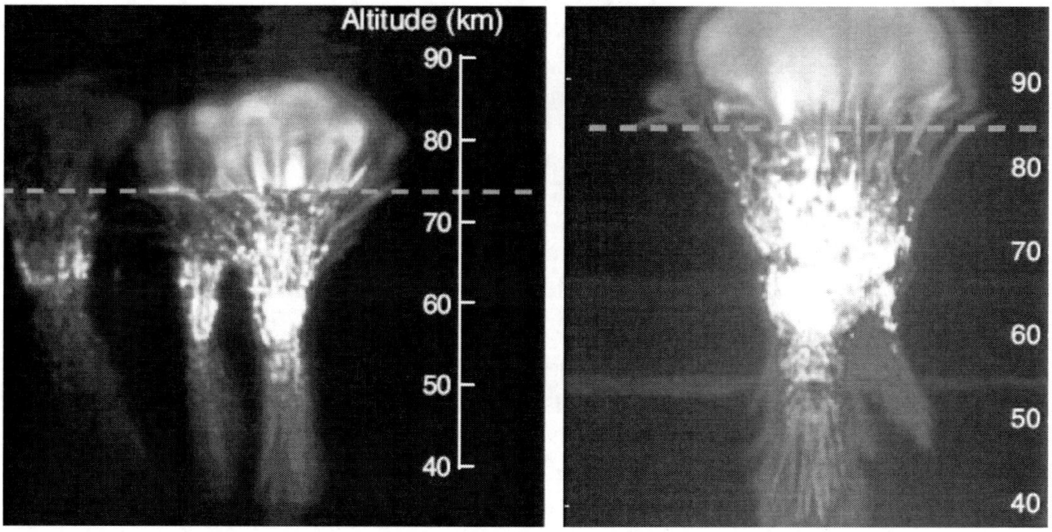

Figure 2. Photo of “Sprites”.

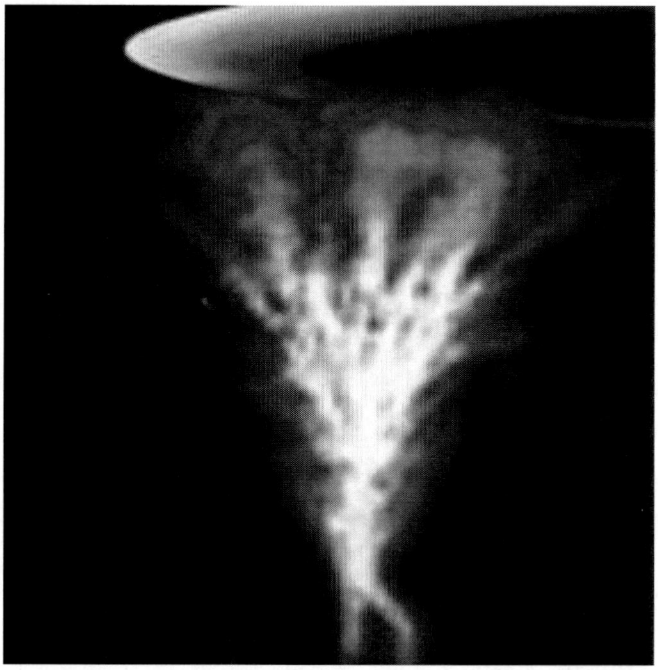

Figure 3. Photo of “Jet”.

ORBITAL ELECTRICAL SOCKET

Electricity creating nicer weather is inseparably connected with thunderstorm electricity. it composes the part of the distributed current outline – the orbital electrical socket (OES) of the atmosphere. As the physical cause for formation OES in the atmosphere serves a sharp increase in the conductivity of air with the height. Near the earth's surface, the conductivity of air is very small and comprises (2-3)$\cdot 10^{-14}$ S/m, which corresponds to the concentration of light ions of approximately 10^3 cm^{-3}. With an increase in the height, because of an increase in the ionization level, determined by 40 km by galactic cosmic rays, and it is above – by the ultraviolet and X-radiation of the sun, conductivity grows almost exponentially with the characteristic scale of 6 km. Already at the height of the lower boundary of the ionosphere (about 80 km) it increases to more than 10 orders in comparison with the troposphere. The conductivity of the earth in the surface layer, and all the water in the ocean also exceeds the conductivity of the boundary layer of the atmosphere by 10-12 orders. Thus, the constantly functioning thunderstorm generators prove to be concentrated in the sufficiently narrow weakly conducting layer between the earth's surface and the ionosphere. Frequently, with the simplified description, the OES of the earth's surface and lower boundary, the ionosphere (70-80 km), they are considered as the facings of one additional gigantic spherical capacitor of the orbital scale, which is discharged in the regions where better weather is charged in the regions of thunderstorm activity. In this case, the quasi-stationary currents of charging are not locked completely to the earth near the thunderstorm clouds, but partially "they are involved" in the superincumbent region of high conductivity and spread on the ionosphere. It is considered that the precise quasi-stationary currents in the first turn "bear responsibility" for the maintenance of a potential difference of ~350 kV between the Ionosphere and the Earth. Since the upper part of the majority of thunderstorm clouds have a positive charge, the potential of the ionosphere also proves positive, and in the regions grow prettier weather the electric field is directed downward, thus causing the conduction currents, which close OES. If the action of generators ceased, a potential difference between the earth's surface and the ionosphere would disappear in not more than 10 min. The orbital electric generator, caused by the non-monolithic nature of the rotation of the plasma sheath of the planet, can serve, together with the thunderstorm generators, the potentially important source of atmospheric electricity. In order to understand the components the mechanism of how OES works and its role in the solar system – terrestrial connections, to be dismantled at three-dimensional – to temporary dynamics, are necessary in the complex experiments, which include the high-precision measurements of electric field and current on the orbital scales and the measurements of the integral activity of thunderstorm generators. To understand working OES, adequate models of thunderstorm generators are necessary, which include the possibility to calculate quasi-stationary and pulse currents, and also the contribution of generators to complete power engineering OES. The results of calculations show that the contribution of mesoscale convective systems in OES can substantially exceed the contribution of single thunderstorms. In spite of the significant disturbance of electric potential in the region of convection, which reaches hundred megavolts, the basic contribution to the complete vertical current and structure of currents in the environment of mesoscale convective systems are determined by region stratifications. Taking into account the latest experimental data and based on them, theoretical estimations created a new description of OES as an open dissipation system, supported by the state of dynamic equilibrium by the entering of energy flow, in the first turn of radiant energy of the sun. It is established that electrical energy is generated predominantly in the regions of the reduced atmospheric pressure and in the zones of cold fronts and comprises on average $5\cdot 10^{13}$ George, two orders higher than energy, concentrated in the orbital spherical capacitor the earth-ionosphere. The average speed of the dissipation of electrical energy $4\cdot 10^{11}$ W exceeds the speed of the dissipation of energy in the tectonic and magmatic processes and characterizes

OES as a completely dynamic system with a very short time for renovation of electrical energy. The process of the transformation of energy in the atmosphere is accompanied by the generation of the dissipative structures of different scales, in the detail of convective systems and aeroelectrical structures, that clearly it is possible to visualize in the form of global processes of the accumulation of electrical energy and its dissipation in the atmosphere. In the near future, the study of power engineering of the atmosphere, structure and dynamics OES will remain one of the vital problems of atmospheric electricity.

CONCLUSION

In conclusion, it is necessary to comment about the interrelation of atmospheric electricity with the formation of the composition of atmosphere and the climate of the planet. Over many years attempts were undertaken to create conducting channels of large length for studying the upper air and solution of special problems. In this connection, the program "Impulsar" is of great interest, which is the combination with the high-voltage high-frequency. Tesla`s source can be useful in the solution to the above problems.

Chapter 12

TO SPACE BY LASER LIGHT

INTRODUCTION

A new approach to the problem of laser jet engine (LJE) is based on the use of shock waves (SW) resonance association mechanism, generated by the optical pulsating discharge (OPD). For the creation, OPD is proposed to use powerful P-P laser radiation with duration of pulses ~ 100-150 ns and high pulse repetition frequency ~ 50-100 kHz. OPD is formed with the help of matrix of reflectors (MR). This is allowed: several times to increase the effectiveness of laser emission use for purposes of LJE creation, to avoid strong impact loads on the apparatus, to exclude the thermal action of laser plasma on the reflector, to decrease the laser emission screening by plasma, to remove the problem of the resonance oscillations appearance into the object moved with the help of laser. In our article, the estimations of possible LJE thrust levels under the conditions of the proposed mechanism of reactive motion are given. Also, the new approaches to the creation of the high power high repetition rate laser systems for the "Impulsar" program are examined.

LJE is the most promising rocket engine of a new class. In our article, the discussion deals with the creation of an automatic spacecraft engine, which will pass its trajectory under the action directed toward it from the Earth's surface, the sequence of high repetition rate short-in-time duration laser pulses.

This is important because LJE is substantially more economical than traditional chemical-propellant engines. In the initial stage of flight, as the working medium adapts to atmospheric air, and beyond the limits of the atmosphere, the insignificant onboard reserve of gas or easily sublimated substance. In this case, specific expenditures for the launch of loads into space can be reduced to 5000-10000 rub/kg, which is approximately 100 times less in comparison with the contemporary scale of world prices. Especially, the possibility of maintaining orbital parameters at the assigned level with the aid of the same laser system, intended for the starting, can be highly evaluated.

In the present time, the study of LJE design possibility is under very intensive realization. Within the framework project "Lightcraft", intensive developments of such systems are conducted in USA, Germany and Japan. So, in November 2000, the American company "Lightcraft Technologies" successfully conducted model tests of a rocket, which rose to the height –70 m during 12.7 s. under the action of exhaust jet, which appears under the effect of a powerful laser emission. In the experiments, the low-frequency 20Hz CO_2-laser with output power 10 kW was used, which in the present time is being upgraded to the level of the output power – 100kW. Mechanical impulse appeared due to ablation of special polymeric material from the concave surface of the reflector, located in the tail end of the rocket, where the laser beam pulses were accepted.

In the beginning of 1973, under scientific guidance of Acad. A. M. Prokhorov in the USSR, work on the study of possibility of LJE design was conducted. The reflector, located in the tail of the rocket prototype, concentrated the obtained radiation in air and ensured the micro-burst that the reactive thrust was created. The successful results of tests of different models of reflectors, which were at the same time as the laser light receivers, had been obtained. One should note that all experiments, pointed out above, were conducted with the use of electric discharge CO_2- lasers with power up to 10 kW, while for the injection into orbit of different highly and technologically effective equipment (global network connections, Internet, photo-monitoring of Earth surface, debris cleaning) radiated power substantially higher is required. Thus, for example, for KA launching with the weight 1000 kg, the laser with power not less than 10-15 MW is necessary. At the present time, this can be the gas-dynamic laser (GDL), since only in this case does the laser technology, in significant measure, intersect with rocket technology, which in 50 years moved sufficiently well in the creation of super-power gas generators and is possible for similar problems. Furthermore, the laser must work in P-P regime with the high frequency repetition of short pulses with the exception of the process of laser emission screening entering by the plasma, which appears with the engine work, for an increase in work effectiveness [1]. In the opinion of specialists – classic missile men – LJE can find effective use in creation of cheap single-stage means of nano- and micro-satellites launch with a mass in the gap 5-50 kg, which according to the opinion of experts of tomorrow, will compose the fundamental basis of a commercial launch program. In the first stage of flight KA at the heights up to 30 km as the working medium in the engine, the usage of atmospheric air is intended, and then to inject satellites into orbit the onboard reserve of the special fuel – ablating substance – in quantities not exceeding 15-20% of the KA started weight is needed.

The experience of powerful lasers creation is collected in the A. M. Prokhorov General Physics Institute of RAS, in the NPO "Energomash" after V. P. Glushko and other organizations of Russia. In OOO "Energomashtechnika" during the last few years, very successful experimental studies of P-P regime in the powerful and well-developed continuous laser systems were conducted. This makes it possible to approach the experimental realization of super-power P-P laser source on the basis of gas-dynamic principle and LJE in the composition of light carrier with control system. The forthcoming complex of works must become a major step toward the future of starting super-lightweight KA low into near Earth orbit. The realization of this project will make it possible to create highly economical LJE reusable, beginning with the removal of payloads of wide designation into space. The key advantage of this new approach is connected with the fact that the source of kinetic energy and payload are untied in the space and launching weight of KA can be lowered to the payload weight only. An even greater K. E. Tsiolkovskii foretold, the starting KA of the future will be achieved with the aid of the electromagnetic waves directed toward the external energy source, laser, which at that time was not known to the World at all. The great interest of scientists and specialists, as confirmed in the last few symposiums in this field of research, was connected with the successful solution in our country of the problem of creating powerful lasers with high repetition rate (50-100 kHz) of pulses of short duration (150-250 ns). This was very recently written "the father of laser reactive motion" Arthur Kantrowits:

> "Dear Victor, thanks for the pleasure of reading your articles from the last symposium on the motion with the aid of laser energy which are now accessible for the scientific community. I already thought more about your works and I count that these remarkable ideas on the quasi-stationary wave, about the light-detonation waves, about the matrix of reflectors those introduced into your new examination are very important for the development of theory and technology of motion with the aid of the light. With great admiration, I do hope for the development of the laser starting in the world and in Russia in particular. I hope that after your publications, the significant activity in the field of laser engine creation and its applications will be stimulated. It would be very interesting to hear further about the progress in this hot field of research. Thanks for your energetic contribution and be healthy.
>
> A. Kantrowits"

High-frequency P-P mode of laser operation is developed and tested on the basis of powerful CO_2 – GDL, Nd YAG with lamp pumping and can be successfully used for other types of the powerful lasers, such as: HF/DF, Nd YAG with LD pumping, COIL and CO. The results of computer modeling for high repetition rate P-P lasers are presented in our review. The data of modeling depends on the characteristics of media, the way of pumping, resonator geometry, and many other parameters of the laser under consideration [2].

COIL	P-P mode starts at frequencies >20 kHz. Depth of modulation -100% starts at frequencies >100 kHz. Pulse duration <250 ns. Ratio: P peak./Paver. = 100 -1000.
HF/DF	P-P mode starts at frequencies >100 kHz. Depth of modulation -100% starts at frequencies >250 kHz. Pulse duration <100 ns. Ratio: P peak./Paver. = 1000 -10000.
Nd YAG	P-P mode starts at frequencies > 4 kHz. Depth of modulation -100% starts at frequencies >40 kHz. Pulse duration <250 ns. Ratio: P peak./Paver. = 100 -1000.
CO	P-P mode starts at frequencies >10 kHz. Depth of modulation -100% starts at frequencies >100 kHz. Pulse duration <250 ns. Ratio: P peak./Paver. = 100 -1000.

At the present time, the project of the GDL with an output power of 20 MW with the varied high-frequency structure of emission is located in the stage of active study in our country. Meaningful results created a notable increase in the effectiveness of laser energy use in the developed mode of emission, which made it possible, according to our estimations, to pass from tens of kg pay-load to hundreds, or even a thousand kg can be obtained. So, one should note that the works in this region, on the basis of their large prospect for different applications, have already begun in Germany, Japan, England, France, China, S. Korea, Brazil and a number of other countries. In this case, practically all participants of these works note the chosen position of GDL as the most promising system from the point of view of its scalability on the basis of the rocket technology to the level of several ten megawatts and other parameters, important for this kind of application.

To the present time, two possible methods of applying the laser emission in the aerospace tasks have been investigated: launching to the space orbit light KA, reduction of aerodynamic drag of the flying bodies, which move in the atmosphere with the high speed were formed. In the problem of developing LJE the generated by P-P laser emission is focused by reflectors near the rear end of KA and periodically, repetitive laser sparks are created. Sparks generate shock waves, which transfer the part of their mechanical pulse to the reflector, located near the tail of the rocket. As characteristic of previous years, the gas-discharge laser systems had demonstrated the pulse repetition frequency, which was limited by the time of gas exchange in the discharge zone in the range of 100 - 300 Hz. For achievement of high average power - 10 MW - for this rate of operation, it should be necessary to use laser pulses with the energy about 100 kJ. With the reduced air pressure, for example for the height more than 15 km, long living plasma sphere, created by each pulse, occupies practically the entire volume of the reflector, which leads to the screening of subsequent pulses into the course ~ of 10 ms. Technical difficulty of that method of energy extraction for the laser also has very strong impact loads with a high energy of pulses. The use of high energy pulses with the small repetition frequency and, consequently, with the very high peak power is also limited by optical breakdown as on the route, so on the surface of reflector. The method proposed by us is the way of overcoming all those difficulties on the indicated basis: usage of laser emission with short duration of pulses and high pulse repetition frequency and mechanism of generated by OPD shock waves resonant association. Under conditions of LJE energy of laser pulse with the short duration (100ns.), effectively (95%) can be absorbed and converted (~30%) into shock waves. In addition to this, it is shown that the specific thrust can be increased several times due to the artificial transformation of radial component of shock waves into the longitudinal one. There are many other advantages of high repetition rate P-P lasers, which should be mentioned here as well [3].

They are:

- Much higher efficiency of energy extraction from media and conversion into laser light, which is important for high aperture lasers scale up;
- Plasma screening effect decreasing due to the decreasing of laser pulses duration and increasing of number of pulses up to a few tens of kHz;
- Decrease of laser beam thermal defocusing due to the optimal temporal structure of laser radiation;
- Localization of energy deposition in space and time and elimination of stressed state of solids under laser action due to replacement of melting by ablation;
- Possibility of unrolled 3d geometry of breakdowns in space due to a very high repetition rate generation of laser pulses, creation of optical, acoustical and electromagnetic fields far away of laser source;
- Much longer distances of optical breakdown conditions for the same optical systems.

Spark Parameters for LJE

Laser emission is focused by the reflector, which can have a form of hemisphere or parabolic. In Figure 1a the significant dimensions of reflectors, focusing on one spark, and also plasma zone, created by the spark are shown. Distance Ff must be slightly between the focal point and the reflector Ff/Rd < 0.2, which follows from the condition of achieving the high value of the specific impulse of thrust. It follows that to avoid the optical breakdown on the reflector and the conditions of transport that the beam must have a large diameter on the reflector. If the intensity of emission exceeds the threshold of optical breakdown, then the plasma front is extended towards the ray, which occurs in heating and the ionization of air as a result of absorbing the laser emission. Since intensity in the sharply focused ray rapidly decreases (shape factor), then already at the small distance from the focus, the disruption of the light-detonation regime of the propagation of plasma front occurs. Further, emission is absorbed for a while in the decay plasma. It follows from the calculations carried out by us that for LJE the greatest interest represents the pulses with the duration ~ of 0.2 s. Maximum energy of laser pulses is limited by the condition for the achievement of high efficiency in the use of laser emission for the thrust application. In the free gas space, the pulse transferred by the shock wave is different from zero only at small distances from the burst center. The same data were also observed in early experiments. For the reflector of parabolic form, the specific impulse is maximum ~ 550 N·s/J at the distance R1, equal with the short duration of the pulse of ~100-200 ns, approximately, the tenth of a dynamic radius, or in another words, the distance, where the value of pulse diminishes is approximately three times bigger. With the increase R1/Rd from ~ 0.1 to 0.33 the value J1 it rapidly decreases ~ by 550-200 N·s/J. B the case LJE the possibility of using the low values R1/Rd is limited by the formation of long-life plasma, whose characteristic radius is compared with the dimensions of reflector. At the last stages of the thermal expansion, a spark is formed in the plasma zone with the low density and the high temperature ~ 8000 k0 ionized gas. The contact boundary of the hot region stops with the pressure balance in the plasma and the surrounding gas. In the approximation of the spherical form of the spark, a radius of plasma region at the pullout moment of pressure makes it possible to obtain the estimation of a radius of plasma formation for the entire operating range of the pressures 0.1-1 atm. The ratio of a radius of plasma zone to a dynamic radius in this case composes the value 0.15-0.25. This relation does not depend on energy and gas pressure. Moreover, its value is located in the same range, where it attains the maximum value of the specific recoil momentum, created by a shock wave. In the reflector, which has the form of hemisphere, the maximum specific impulse of the thrust reaches $R_1/R_d \approx 1$. Let us give the values of a radius of plasma formation and time of its formation for energy 10^5 J and the pressures of atmosphere 1 and 0.1 atm. The parameters correspond: peak pulse power – $2 \cdot 10^7$ W and to the pulse repetition frequency – 100 Hz, and also to start and to the end of the stage of acceleration KA in the regime LJE. With pressure 1 atm. the size of the plasma sphere will be 25 cm. With the time of formation on the order of 1 ms, while at a pressure of the atmosphere – 0.1 atm. the size of the plasma sphere will already comprise 50 cm with the time of formation – 2 ms. Cooling laser plasma occurs due to turbulent mixing with the cold surrounding gas. The characteristic time of this process more than to the order exceeds the time of its formation. Here the spark with spherical form has been examined. On the matter itself, the spark in LJE has a form of a cone with a large apex angle. This worsens the situation. With the specified conditions at the last stage of the expansion of spark, a cumulative jet can be formed, in which the gas moves in the direction of the reflector. Plasma ceases to be singly connected, taking the form similar to the torus. Thus, the use of a regime with the maximum pulse, presented in the early work of many authors, will be accompanied by the contact of plasma with the surface of the reflector, which, undoubtedly, will lead to its destruction.

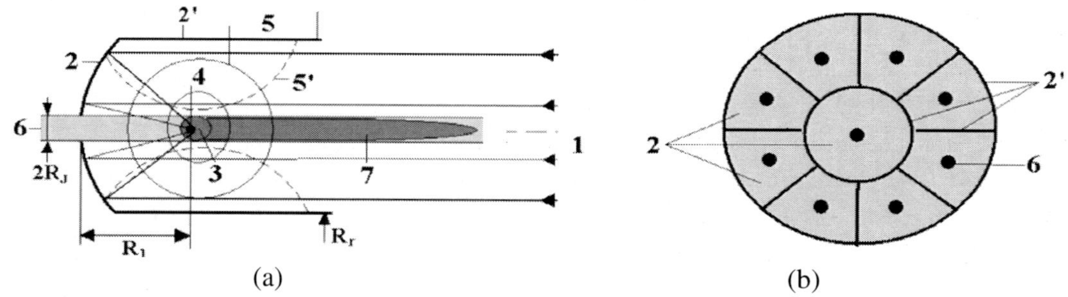

1 – high rep. rate P-P laser radiation,
2 – back side of reflector-zone of laser pulses interaction,
2' – side wall of reflector, 3 – cavern, 4 – OPD, 5 – Shock wave,
5' – reflected shock wave, 6 – gas stream, 7 – plasma stream.

Figure 1. Scheme of reflector (a) and possible structure of matrix of reflectors (b) for LJE.

Mechanism of Shock Waves Resonance Association

The effect of resonance association separates shock waves into the low-frequency quasi-stationary wave (QSW) in the general case consists in following. In continuous medium are created consecutive periodic disturbances – shock waves, whose initial velocity is more than the speed of sound. If the speed of the displacement of the field of pulsations is lower than the speed of sound in the medium, then shock waves are united and are created BY SQ. Depending on the time-spatial structure of pulsations, the mechanism is manifested in the form the effects, the basic property of which is the large length of the high-pressure area. The mechanism of shock waves association does not contain limitations to the type of medium and source of pulsations, to its energy. Depending on the structure of pulsations QSW can have various forms.

The realization of mechanism kV makes it possible to remove the problems examined above: the screening of emission and the thermal action of laser plasma in LJE. Furthermore, QSW makes it possible to substantially increase the effectiveness of the use of laser emission in the calculation of an increase in the specific thrusts by the unit of power. Let us determine requirements for energy and pulse repetition frequency of P-P of the laser emission, which satisfies the solution of this problem. In this case, we can examine two methods, based on the use of "spherical" and "flat" OPD. In both cases, a flat short circuit is formed, the calculation of the use of both the geometry OPD and the geometry of reflector, and also the organization of the process of the introduction of energy of laser emission into the engine.

LJE on the Basis of Shock Waves Resonance Association for the Case of MR

In the general case, MR is represented in Figure 1b. The problem of developing LJE on the basis of shock waves merging and P-P of laser emission with high (to 100 kHz) pulse repetition frequency, and also flight path control aid, LJE was decided for use with MR. The schematic of the engine is with its collection from N mono-reflectors. P-P brings the emission to MR with the energy of the pulses q with the average power W_C. Generally, the elements of MR are identical. MR creates matrix OPD, each of which is stabilized by the air jet of the incident flow. A quantity of matrix elements in the general case determines the high-speed regime of gas jet. In our case, a quantity of elements in the matrix was $N = 8$.

In our calculation, the power of laser emission was taken as the equal of 20 MW, the pulse repetition rate – 100 kHz, energy of each pulse in this case equaled 200 J. Calculations showed that for the accepted examination parameters in the case of normal atmospheric pressure, the complete engine thrust will equal 100 thousand N, and at a pressure of the order of the tenth atmosphere its value will fall to the value – 36 thousand N.

The optimum gas flow speed for both cases proved to be equal to 2.5 km/s and 5.4 km/s, respectively. One should also note that flight control is possible with the aid of the laser engine with the calculation of a change in the thrust in the equivalent components MR. However, an increase in the number of elements of MR is accompanied by a significant increase in aerodynamic drag on the air ducts, and the selected number of elements is close to the optimum.

In the experiments, the model of a rocket with the laser engine had a cylinder with diameter ~ 8 cm, length ~ 26 cm and weight 1.1 kg, suspended to four thin wires with total length of 1.1 m. Construction allowed displacement only in the axial direction. On the chamber end is fixed the reflector – change cylindrical cap. Laser emission was introduced into the camera through the lens with a focal length of 17 cm. The jet of argon was formed with the expiration from the high-pressure chamber through the opening with a diameter of ~ 3 or 4 mm. The speed was regulated by the pressure of argon, which was given into the camera on the flexible hose. The force, created by jet and shock waves, was transferred with the aid of the thin (~ 0.2 mm) molybdenum wire to the load, which stands on the strain scale weights.

OPD burnt in the flow, which was formed with the outflow of gas from the camera through the opening in the reflector. OPD was created by P-P laser emission – $f = 50$ kHz and 100 kHz, in this case the average radiated power comprised $P \approx 1200$ W.

Thrusts with f = 50 kHz, The V = 300 m/c proved to be equal – 40 g., and with The V = 400 m/c – 69 g. The value of the thrust coupling coefficient proved to be equal – 1.06 N/kW. The experiment confirmed that P-P laser emission creates a stationary thrust with very high coupling coefficient – 1000 N/MW. A difference between the experimental results and theoretical estimations, whose value >4500N/MW, is connected with the need for further optimization of the process of the introduction of laser energy into the plasma OPD.

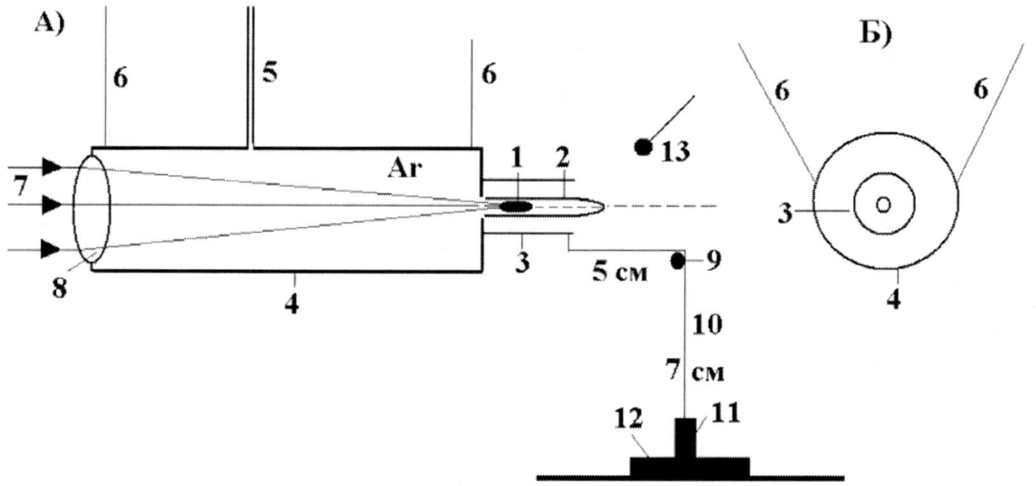

Figure 2. Experimental scheme.

RESONANCE PROPERTIES OF SYSTEM "LASER - KA"

An important parameter in the estimation of the resonance properties of system "Laser-KA" is the ratio of the duration of laser pulse to the transit time of sound throughout the entire length of the rocket. In the practical sense, there is great interest in the case when the uniform load, under the action of laser pulse, is created throughout the entire length of rocket. If the pulse has a very short duration, then acceleration is non-stationary, the length of that excited in KA waves are much less than the length of rocket. The case, when the pulse duration and the transit time of acoustic wave are compared, corresponds to the resonant step-up of waves.

Let us make the resonance properties of the system for the laboratory experiences in the case of KA with the mass – 20kg, the length of the rocket of equal – 200 cm, acceleration – 100 m/s^2. Average power of P-P radiation – 4 MW, the energy of pulses generated with repetition rate – 200 Hz is equal – 20 kJ. Calculations show that for small pulse repetition rate of P-P laser radiation, the extremely strong dynamic structural loads of KA accompany the acceleration mode. They have a resonance nature, since the wavelength of compression is compared with the length KA. Furthermore, with an increase in the length KA of about 4 m and the repetition frequency of the laser pulses up to – 1 kHz, the natural frequency of oscillation of a rocket is equal to the pulse repetition rate of the laser and we obtain the fulfillment of conditions of resonance. This is very dangerous, since resonance loads can exceed the safety factor of the rocket, and it will be destroyed in the flight.

Thus, estimations show that the thermal contact of plasma with the reflector and strong dynamic loads are inevitable with the low pulse repetition frequency. The situation is aggravated by the excitation of resonance oscillations in the KA. These difficulties are surmounted only with the use of a method, based on the association of shock waves. Calculations and experiments confirmed the possibility of significant stationary thrusts designed with the use P-P of emission with high pulse repetition frequency.

LASERS FOR THE "IMPULSAR" PROGRAM

Since the creation of the first lasers at the beginning of the 1960's, the search is continuously conducted for new principles of construction of the sources of coherent optical radiation, which expand the region of their application both in scientific and applied research, and in military equipment. In this case, as a rule, they approach an increase in the energy of the unit pulse with the simultaneous shortening of its duration and for an increase in the average power of optical radiation in the calculation of the periodicity of the process of its generation.

To the present period, in the lasers of continuous and P-P of the regime of the generation of emission with the frequencies of 200-300Hz the output power10 kW – for the solid-state lasers and to 1MW – is achieved in gas lasers. The cross section of the bundles of such lasers comprises from ones (for the solid-state lasers) to tens of centimeters (for the gas lasers). These are bulky installations with complex systems of pumping active medium, heat withdrawal, and formation of output emission. In this case, the basic obstacle for creating the super-power and compact lasers is the problem of the heat removal from the active medium. Even in the case of CO_2 lasers, which have relatively high efficiency (to 15%), it is necessary to ensure the effective draining of heat from the entire volume of the active medium. In the contrary case, overheating the gas mixture will lead to the disturbance of its uniformity and as a consequence, worsening the quality of the output emission. This problem even remains in solid-state lasers, because of the impossibility of the conclusion strictly of active medium from the zone of pumping and cooling in the auxiliary volumes, even using swiftly developing recent sources of the selective pumping of solid-state lasers, laser diode rules and matrices, which ensure efficiency. Solid-state laser, even at the 50% level, does not remove this problem. In powerful lasers, the heat emission in the active elements remains very high, which leads to a significant reduction in the optical quality of laser emission.

The creation of powerful lasers with a large integral cross section of output bundle on the basis of the amplifier channels, which use geometry of slabs, represents a complex scientific and technical problem. Multi-channel solid-state lasers with the level of output power into several ten kilowatts are created and will be presently tested in the USA and Japan. These are very complicated physical-technical devices, which do not practically function without constant technical support. The attempt to create solid-state lasers of high power for the calculation of an increase in the cross-section of an active medium with the simultaneous decrease of its sizes along the axis of a resonator, the so-called "disk lasers", led to new technical problems, connected with the suppression of strengthening the spontaneous emission of the pumped active medium in directions perpendicular to the optical axis. Spontaneous emission leads to the undesirable removal of the inversion of active medium and, respectively, to the decrease of the effectiveness of laser.

From other side the solid-state lasers with the diode pumping, built according to traditional optical diagrams, which have output power on the order of 3-5 kW, are already sufficiently ordinary instruments in the West. The existing technical and technological solutions make these lasers possible by requiring maintenance of the limits in the entire period of operation, which composes tens of thousands of hours.

It is obvious that one of the effective solutions of creating high-power solid-state lasers consists of the multi-channel principle of their construction. Laser, in this case, is "collected" in the form the lattice of identical channels, each one has a laser-emitting generation. This lattice of lasers forms an optical source with composite aperture, whose output emission is the totality of the beams of separate lasers. The small cross section of channels makes it possible to effectively and simply accomplish the pumping of energy into the active medium and heat removal from its volume. According to this diagram, it is built with a 19 - kilowatt laser made by the firm "Northrop Grumman" [4].

Before the appearance of diode pumping, the construction of solid-state lasers according to this diagram was problematic, because of the practically unsolved technical problems of cooling the tightly packed lamps and active elements. Further improvement in the output parameters of multi-channel laser systems is possible due to the significant decrease of the optical thickness of the amplifier and, consequently, the passage to the disk geometry of amplifier channels.

In multi-channel type lasers, the problem of obtaining the high directivity of emissions is effectively solved by the methods of phasing. Actually, the angular divergence of the emission $\Theta = \lambda/D$, where D – the diameter of aperture. In spite of that, the aperture of the lattice of lasers to the order exceeds the aperture of separate emitter, the divergence of emission is determined by the divergence of beam of separate emitter, and, respectively, the intensity of the emission of multi-channel laser is determined by the expression:

$$I = N\, I_0$$

Where N - the number of emitters;

The I_0 - the intensity, created by one channel.

In the case of the non-phased emission, each channel emits a wave in the arbitrary phase, which does not depend on the phase of the other channels. If we make it necessary to emit lasers in this phase, then the picture changes radically. The divergence of emission will be determined by the distance between maximally diverse lasers, and the intensity will be determined by the expression:

$$I = N^2\, I_0$$

Specifically, according to this diagram of the developers of firm "Northrop Grumman", it was substantially possible to move in the plan of the integral divergence of the united laser beam -1,73 of diffraction limit for the composite aperture with 30% of optical efficiency. At the present time, it is possible to go considerably further. Authors of the solid-state laser breakthrough have obtained and published results at the level of 105.5 kW. It is obvious that a quantity of channels with specified conditions can be significantly increased and output power on the level of several hundred kilowatts and even bigger is not limited. It is important to note that with this line-up of laser system, it is sufficiently easy to vary the pulse duration and on the duty cycle P-P the regime of the work of the entire system as a whole.

Works in this direction have been conducted in the Prokhorov General Physics Institute RAS since 1987. Experimental results have been obtained in the study of the disk geometry of laser emitters, in the development of the methods of control, and monitoring of the parameters of powerful laser systems inside the resonator, on the phasing of laser diode structures. So, for example, it was possible to obtain the phased generation of the large number of separate elements of laser diode arrays, and in turn, the phasing of the matrix of laser diodes.

In the example of realization of the scaled system of the phased multi-channel solid-state laser, it is possible to examine the structure of solid-state lasers with the diode pumping on the basis of rectangular elements of the "chessboard" type, which ensures a maximally tight packing of separate laser sources. This packing makes it possible to maximally decrease the overall sizes of assembly, which is principally important for the subsequent synchronization of sources. In this case, very effective heat withdrawal from the volume of active element - because of its small thickness and the developed cooling surface. In this configuration, the power capacity will not be limited by heat withdrawal, but by the possibilities of laser diode rules. The represented system makes it possible to obtain the real apertures of mirrors and phasing optical system in the limits of many tens of centimeters. The production of the optical elements of the larger aperture is not problematic. In this case, it is sufficient to realize the basis of the existing element base to obtain the level of the output power of 500-1000 W from the separate emitter. This makes it possible to speak about the possibility of designing of compact laser source with the diffraction quality of emission at the level of MW. Furthermore, in the configuration indicated, it will be simple enough to obtain the regime of the q-switching and, therefore, periodically repetitive sequence of pulses with the large peak power. As a result of the fact that with the phasing, each of the emitters exchange energy with their neighbor, and obtaining the regime of modulation is sufficient to modulate the emission of a separate source, for example with the aid of electro-optical shutter. In this case, locking the minimum aperture is required, which makes it possible to avoid the technical problems, characteristic of the large apertures of emitters. From the general physical considerations, it is possible to count that with the discussed level of power and the required value of the beam strength of optical elements in this development by phased multi-channel high-frequency P-P of laser system with the semiconductor pumping can be achieved the quality of the laser emission at the level of M2=1.2 by the force of the weak dependence of the divergence of the phased multi-channel disk laser system with the semiconductor pumping on the number of channels, and this follows from the ideology of multi-channel systems and our experimental work experience, further increase in the number of channels of laser system to the level of output power in several hundred kilowatts the real quality of summary laser beam for both cases (continuous and P-P) will change insignificantly and it will remain within the limits of M2=1.5.

High-frequency P-P the regime of generation with the varied time structure of the emission of the phased multi-channel laser system with the semiconductor pumping will contribute to overcoming the problem of the plasma screening of target, increasing the power output in the active medium, to improvement in the quality of laser emission, to increase in the maximum parameters of the effective action L p-p/L cw. = (I p-p/I cw.)$^{1/2}$, to the expansion of the contemporary technology applications spectrum. Created in the time of fulfillment, R & D laser systems will make it possible to conduct the number of purposeful studies, necessary for comprehension of the significant advantages of high-frequency P-P of a regime for a number of new technologies. In particular, the created scientific and technical experimental base will make it possible to estimate powerful high-frequency P-P lasers, and their effectiveness for solving the following problems [5]:

- Ecological applications of the powerful lasers: extinguishing fires on natural gas towers, express cleaning of aqueous surface from petroleum-containing products after the wrecks of vessels, cleaning the extensive and complex surfaces from the pollution, including radioactive;

- Protection of valuable and ecologically dangerous objects from lightning strokes;
- Extraction of minerals under the special conditions (gas condensate from the large depths of seas and from the permafrost, boring rock layers, etc.;
- Unique machine building (treatment of super-hard materials, the creation of the three-dimensional prototypes of objects with separately complex forms, etc.);
- Energy transfer up to the significant distances with the aid of powerful lasers;
- Obtaining powerful plasma radiation source in the range of tens-of-hundreds of angstroms.
- Realization of the arbitrary geometry of breakdown in the three-dimensional space.

CONCLUSION

In conclusion of the material presented, one should say that as a result of the conducted investigations, LJE may aid in the solution of the following interesting problems: creation of the interceptor of manmade space debris and other dangerous space objects, such as asteroids, comets, meteorites and so on, launch of micro-objects with the super-high acceleration to the space, realization of orbital scale conducting channels for energy delivery from space to the ground and vice-versa.

REFERENCES

[1] V.V. Apollonov, Yu.S. Vagin *et al,* Patent Russian Federation № 2175159.

[2] V.V. Apollonov, High power lasers for "Lightcraft" applications, Proceedings of ISBEP-6. Santa Fe, (NM), USA, 2009.

[3] V.V. Apollonov, Yu.S.Vagin et al, High-frequency pulse-periodic mode for high-power laser, Proc. of ISBEP-2, Sendai, Japan,2003.

[4] 4. P. Rincon, "Record power for military lasers", Science reporter, BBC News, 2010.

[5] V.V.Apollonov, "Vpered k Tciolkovskomu", Moscow, "RT Soft" 2010.

Chapter 13

LONG CHANNEL FOR ENERGY TRANSFER IN SPACE

LIGHTNING PROTECTION SYSTEM

At present, the development of new methods of the single-wire, long-range energy transfer is of great scientific interest. Modern methods are a continuation of studies performed by Tesla at the beginning of the XX century. In particular, not so long ago a highly effective energy transfer between two resonant circuits connected by very thin (100 μm), 1-km-long wire was demonstrated. In these experiments, use was made of Tesla-type transformers with an output voltage of about several ten kilovolts and an output power up to 10 kW [1]. An effective replacement of wire by a laser-plasma channel in this configuration of the experiment seems to be an obvious improvement of the demonstrated technology. On the other hand, in the last 20 years many laboratories of the world have carried out interesting studies aimed at creation of effective laser lightning protection systems. As is known, a pulsed laser can produce long enough laser sparks, which can connect a thunderstorm cloud with a grounded iron core, i.e., a classical lightning rod. An electrical discharge channel with a maximum length of 16 m has been produced in Russia and Japan using a pulsed CO2 laser with an energy up to 0.5 kJ, which was developed and fabricated at the A.M. Prokhorov GPI, RAS.

Later, we also demonstrated much longer laser sparks obtained with the aid of conical optics, which offers much more advanced characteristics from the point of view of plasma channel conductivity and very short time delay of its effective start. We were the first to outline the possibility of using (instead of a single wire) long laser sparks produced by conical optics to connect two resonant circuits.

At the present time, among an enormous quantity of different types of lasers for electrical discharges directed by the wide spectrum of laser spark application, only two type of lasers are found as the most effective for this particular task: pulsed sub-microsecond CO_2 lasers and femtosecond (FS) solid state lasers.

Possibility of one or several parallel highly ionized channels creation, so called filaments, with a characteristic diameter of 100 mkm. in the atmosphere along the direction of the laser beam propagation is the major feature of FS lasers. With estimated value of electron densities of such filaments (up to 10^{16} 1/cm$^{3)}$ and with wavelength of the laser in the interval 0.5 - 1.0 mkm plasma, practically, does not absorb laser emission. In this case, the length of channel is determined by the intensity of laser emission and in the case of energy of the FS pulse ~ of 100 mJ can reach the length up to 150 m. The possibility of a considerable increase of filaments length is connected with the application of unique super powerful FS laser systems with energy about 1 J and even more. Such a system in our days looks like a unique laboratory setup and can hardly be used for commercial and other applications. Another limiting factor is a very high resistance of conducting channel based on FS laser filaments. Characteristic value of resistance of such a channel is on the level of 0,1 Mohm/m [2].

In the case of pulsed CO_2 laser use in combination with classical spherical or conical optics, the length of spark channel is determined by the parameters the caustics of laser and that its part in which the intensity of emission exceeds the threshold for the optical breakdown of air. The presence of intensity threshold is connected with the special features of the mechanism of plasma formation where the avalanche of multiple initial "background" electrons in the field of laser emission takes place. A vital difference in the geometry of spark in the case of long pulse and spherical optics from the case with the FS lasers is the fact that the laser spark occupies volume inside the caustic curve from the focal plane to the side of laser, since the laser radiation cannot traverse self created plasma shield. In this case, for the super long spark formation, it is important to use CO_2 laser pulses with traditional form, i.e., with short (50-100 ns) spike and long (1-3 mks) tail [3]. Long tail ensures the additional heating of laser plasma, moving along with the laser beam by expanding to the caustic and, respectively, increasing the length of laser spark. The length of laser spark in several hundred meters has been obtained even in the USSR with the aid CO_2 laser with output energy ~ of 5 kJ and the classical form of the pulse.

Figure 1. TC-300. Internal view of an electrical discharge CO2 laser with an output of 300J.

However, conical optics use the possibility of laser plasma channels formation of substantially larger length does appear. According to our estimations the length of such channels in the case of conical optics in combination with pulse-periodic (P-P) CO_2 lasers with an average power of 0.1 – 1 MW can reach several kilometers. In this case, the geometry of spark is completely different and the region "of collapse" is determined by the diameter of the beam on the surface of conical mirror only and by the angle of taper. Spherical optics plasma front moves in the direction of laser beam, i.e., from the surface of the focusing mirror in outside, in this case different spatial domains in the section of laser beam form different sections of plasma cylinder. Therefore the formation of very long laser sparks even with comparatively short lengths of laser pulses is possible. The photo registration of spark showed that it consists of the set of the almost contiguous between themselves elements, on the form resembling "Chinese cap" and located on the identical distance from each other.

Thus, for the experiments conducting the electrical energy transfer between two resonant circuits, situated on the distance of not more than 100 m, it is possible to use both types of lasers. At the same time for commercial energy transmission system creation, the powerful high repetition rate P-P CO_2 lasers appear more promising, since they possess the substantially lower threshold of the optical breakdown in air and are technically more advanced in scaling up of output power to many hundreds of kilowatts.

As has already been discussed above, the geometry of plasma cylinders for two types of lasers was completely different: the collection of thin filaments - FS laser or cellular single channel with more or less uniform plasma – powerful sub-microsecond CO_2. At the present time, it is possible to assert that powerful pulse CO_2 laser and the mechanism of laser radiation breakdown of gas medium have decisive advantages over the FS ideology of the energy transfer by long lengths laser plasma channel.

First experiments carried out by our group with pulsed CO_2 laser having the traditional form of lasing in the beginning of 90th were very much informative. By potential applied to "the ends" of laser spark (up to 600V) we showed that the glow of laser plasma remains not more than 10 mks, at the same time the conductivity of laser plasma channel exists considerably longer – up to 100 mks. High repetition rate P-P laser emission is capable to solve the tasks of the reproduction of the conducting channel during a long period of time and to increase the conducting channel length significantly. The most essential drawback of the described method is the fact that it cannot be directly used in vacuum, which means beyond the limits of the Earth's atmosphere.

New method of the energy transfer and new approach to the transmitting channel formation are directed toward essential simplification of electrical energy transfer methods, at the same time they can be effectively applied to the non-conducting media conditions. Necessary results can be achieved due to the "Impulsar" technology.

The energy transfer in this case is achieved because of the fact that in the construction of the focusing system the source of the medium creation in the form of easily ionized substance, which forms the steam-gas medium with low threshold of the breakdown, which is especially important in the case of insufficiently high medium pressure or directly in the vacuum.

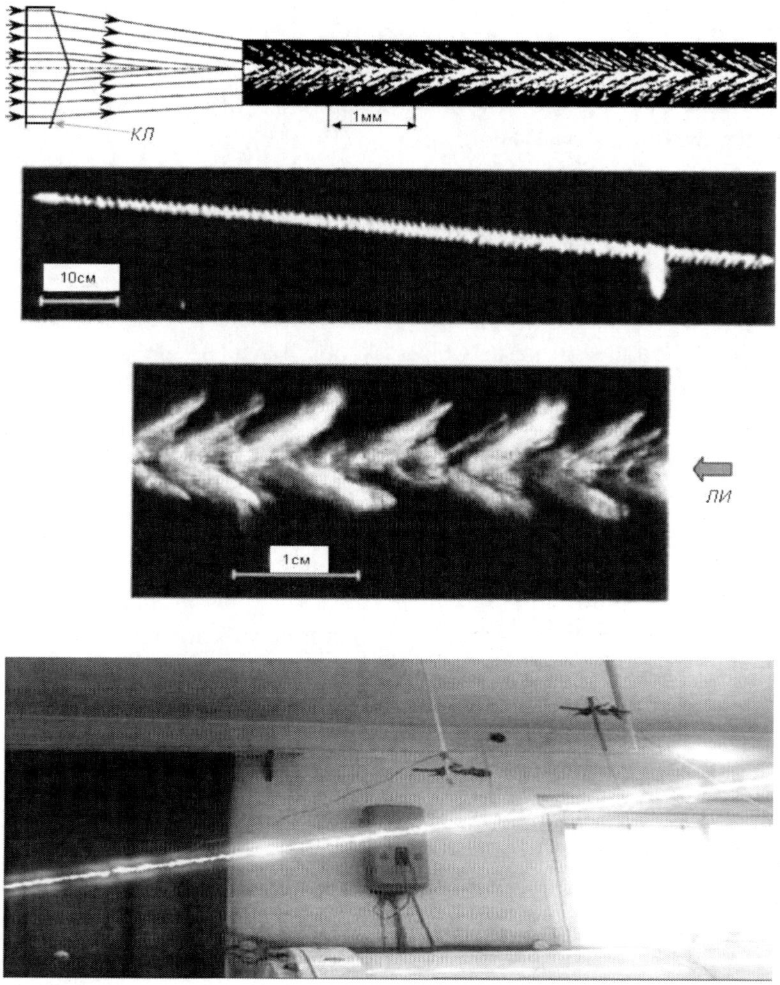

Figure 2. Long conductive channels in atmosphere and their structure.

New Approaches for Conductive Channel Creation

The displacement of the optical focusing system in the media and interaction of P-P laser radiation with energy in the focus of optical system sufficient for the breakdown ensures the formation of the continuous current-conducting channel in the air medium due to its ionization, in this case. As calculations show, the frequency band of laser pulses ensures the continuity of the formed current-conducting channel in the appropriate speed range of the moving focusing system, which can be realized in this particular medium. Actually, each pulse of the laser, with the help of focusing system can create a certain extensive region of plasma, which applies to the relatively small section of space in the trajectory of the displacements of that focusing system. If the repetition rate of pulses follow each other with small frequency, the displacements of the plasma regions after focusing system will be something more similar to the dotted line. However, with higher frequency, for example more than 10 kHz, and optimum for the medium speed the displacements of ionization regions will no longer have breaks and the formed by this process conducting channel will be continuous.

However, after entering of moving optical system into much less density layers of the atmosphere and further into the vacuum, the problem of the scarcity of medium appears, which could complete a deficiency in the ionized gas both as the source of conducting medium and the means of shock wave accumulation and transfer to the object of the motion of necessary quantity of energy. For this case the moving optical system must be supplied with the source of medium creation in the environments of laser focus in the form of sublimated under the action of incident laser radiation easily ionized substance, containing the nano-particles of metal or substance, which under the action of laser emission ensures the synthesis of such a nano-particles. The presence of easily ionized substance source, which due to the sublimation process under the action of laser emission, makes it possible to form the current-conducting channel in the vacuum ensures the continuity of the canal in this case. Presence of nano-particles of metal in the sublimated substance allows simultaneously with improvement of channel conductivity conditions to increase the value of the specific impulse of thrust of laser jet engine (LJE), containing mobile optical system and reservoir with the sublimated substance, which in the essence with this use has determined the time of reaching the required length of electrical power transmission channel [4].

The introduction of high voltage electrode into the current-conducting channel, formed by the moving optical focusing system, allows, from one side, to stabilize the conducting channel of the required length and direction and from the other side to transfer energy by this channel very effectively. In order to facilitate the breakdown conditions of the medium in the optical system focal point with the appearance of not conducting channel only but high energy shock wave as well, it is expedient to use the easily ionized substances, which form the steam-gas medium as the source of medium creation with low breakdown threshold. The latter fact makes it possible to decrease the requirements for the laser pulse breaking energy value.

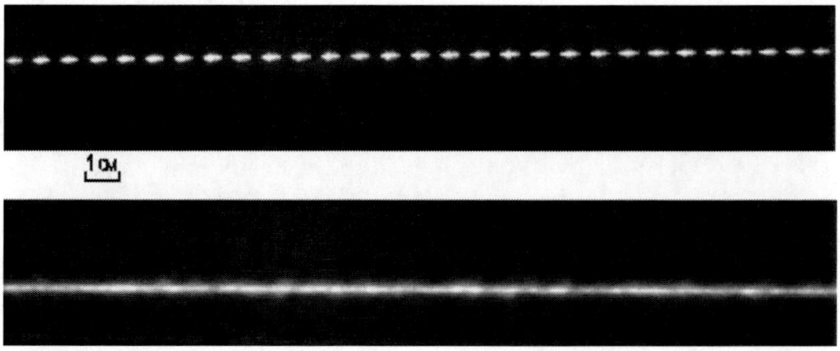

Figure 3. Time controlled evolution of trains in the space for F = 20Hz, 50 cm/s and for F = 25 kHz, 250 cm/s correspondingly.

The moving focusing system can be executed in the form of the conical body or the combination of cylinder and cone, inverted by apex to the side direction of its motion. In the end of the moving system, special optics is established, which can be selected from the number of known optical systems, and ensuring the focusing of the laser light, which falls down to the optical system at the certain distance from the surface. In the body of moving system the sublimated under the action of laser emission easily ionized substance is placed, which contains the nano-particles of metal or substance ensuring their synthesis. As the easily ionized substance wax, paraffin, delrin (high-quality acetate homopolymer) and other polymeric substances can be used. The nano-particles of metal can be of significant dimensions, for example 10-100 nm and can be prepared from the metals such as aluminum, tin, copper, Na, K, Cs, lead and so on. The carbon-containing substances in the combination with the alkali metals can be used as the substance, which makes it possible to synthesize the nano-particles of metal or the metal-composite materials, which possess sufficiently high electrical conductivity. From the literature they are well known – fullerenes on the basis of alkali metals, fullerenes on the base of coppers, which demonstrate high-temperature superconductors properties with tc >of 140K. For obtaining the nano-crystalline powder the plasma and laser methods of heating can be used. Thus, the nano-particles of carbides, oxides and nitrides with the aid of the pulse laser heating of metals in the rarefied atmosphere of methane (in the case of carbides), oxygen (in the case of oxides), nitrogen or ammonia (in the case of nitrides) were obtained. The pulse laser evaporation of metals in the atmosphere of inert gas (He or Ar) and gas- reagent (O2, N2, NH3, CH4) makes it possible to obtain the mixtures of the nano-crystalline oxides of different metals, oxides, nitrides or carbide – nitride mixtures. Composition and the size of nano-particles can be controlled by change in pressure and composition of inert gas atmosphere and gas reagent, by power of laser pulse, temperature gradient during the cooling process. Two reasons for the fine powder to be more conductive can be taken into account. One of them is the strength of the electric field increase due to the size and shape of the particles. Another one is the concentration of particles in the dust plasma track of "Impulsar" [5]. The content of nano-particles in the sublimated substance usually composes 10-15% of the mass. As the source of laser emission, high repetition rate P-P laser should be used. The moving focusing system is started directly from the earth's surface with the aid of high repetition rate P-P laser emission with the energy sufficient for the medium breakdown in the focus of optical system.

Because of laser radiation concentration in the relatively small volume, the breakdown of medium takes place. Air breakdown appears in the region of ionization, which applies to a certain distance following the moving focusing system, and the shock wave, which gives to a moving system additional momentum, leading to its acceleration. Upon the transfer of moving focusing system into rarefied layers of the atmosphere or open space, breakdown and appearing in this case shock wave will be achieved in the easily ionized substance, which the moving system has been supplied in advance. With the arrival of the moving focusing system up to the assigned distance or on reaching by the current conducting channel the energy receiver it is necessary to place the electrode of high voltage source into the canal and to ensure the energy transfer process. The conductivity of channel proved to be sufficient for the discharge of capacitor bank to the ground just after reaching by the moving optical set of the grounded electrode, located at the final point of lift.

The results of dust plasma electrical breakdown tests are presented here without detailed explanation of physical matter of experimental arrangements. The most simple and effective dust plasma production methods were involved into our investigation. Best result of specific voltage for dust plasma breakdown was measured for electrical Cu wire explosion approach. The results of dust plasma modeling are [6]:

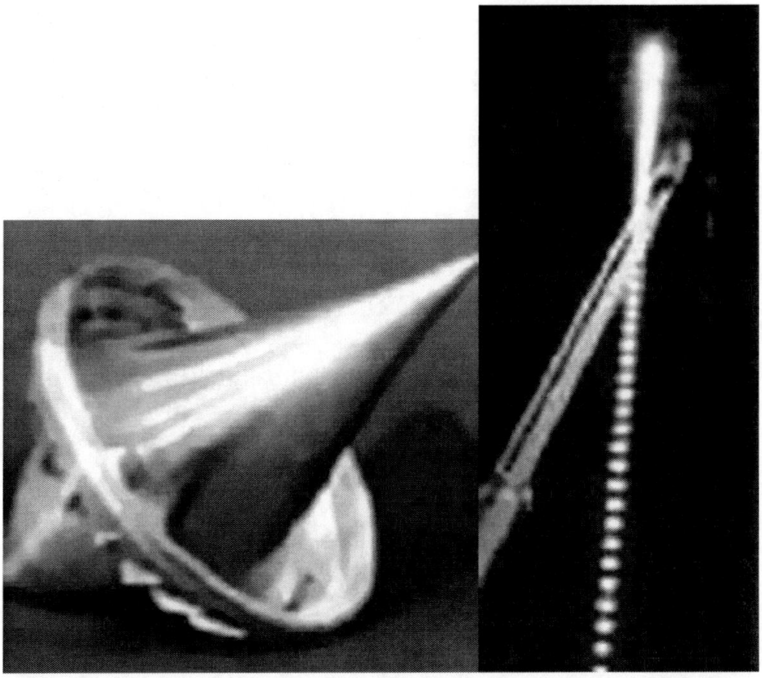

Figure 4. Launch of vehicle by low rate P-P laser (Rep. rate -20Hz, P=10kW, White Sands, NM).

- P-P CO2 laser with train of pulses and total energy-20J. Dust particles size 50-250 nm and concentration N=10 to the 10-12 power.
- For dust plasma produced by laser ablation: paraffin wax + graphite or permalloy powder -300v/cm;
- For nano-particles of Al, Fe, segnetoelectrics -350 v/cm;
- CO2 single pulse laser channel, produced by conical optics - 100 ohm/m, 100 V/cm;
- Long channel (filaments) produced by solid state laser - 0,1 Mohm/m, 7,5 kV/cm;
- Long channel produced by wire explosion - 10 ohm/m, 45-50 V/cm;
- Data for wire explosion channel, corresponding to the delays: 10 mks -50 mks – 75 mks; 97 V/cm-124V/cm-166V/cm.

It should be mentioned here that min. resistance value very much depends on cross-section of dust particles sheaf, concentration and size of particles, repetition rate of laser pulses, average power of laser, gas pressure, humidity and many other parameters.

Conclusion

Powerful pulsed lasers are capable of creating large scale current conducting channels, which can be located at the arbitrary distances from the source of radiation. Channels with the length of about a hundred meters can be achieved with low energy single laser pulse. Beginning from the 70's, successful attempts of the problems: interception of lightning bolts and blocking of over voltage waves on the electric power lines were undertaken.

Successful developments of high repetition rate P-P powerful lasers technology and the technology of the “Impulsar” system make it possible to foresee the realization of well conducting channels with length up to several ten and hundred of kilometers for the purpose of energy transfer for significant distances, creation of new and promising systems for mastering outer space power engineering, and motivation for significant contributions and essential improvements of the global ecology of our planet.

REFERENCES

[1] V. V. Apollonov, V. R. Sorochenko, Yu. A. Shakir, “Experimental investigation of resonance single wire line of energy transfer” Report of GPI RAS, 2009.

[2] “Teramobile: Site in INTERNET, French (CNRS)-German (DFG) scientific research program funded by ANR: “*First mobile terawatt laser in the world for atmospheric studies*.”

[3] V.V. Apollonov, “To the space by laser light”, Vestnik of RANS 1, (2008).

[4] V.V. Apollonov, Yu.S. Vagin et al, “*Conducting channel creation in a non-conductive media”,* Patent № 2009118874, 2009.

[5] V.V. Apollonov, N.V. Pletnev, *“Special features of conducting channel formation in atmosphere*”, In press.

[6] V.V. Apollonov, N.V. Pletnev, “Formation of extended conducting channels in atmosphere”, QE 42(2), 2012.

Chapter 14

HIGH POWER OPTICS FOR HIGH POWER/ENERGY LASERS

The advent of the laser has placed stringent requirements on the fabrication, performance and quality of optical elements employed within systems for most practical applications. Their high power performance is generally governed by three distinct steps, firstly the absorption of incident optical radiation (governed primarily by various absorption mechanisms); secondly, followed by a temperature increase and response governed primarily by thermal properties and finally the elements thermo-optical and thermomechanical response, e.g., distortion, stress birefringenous fracture, etc. All of which needs to be understood in the design of efficient, compact, reliable and useful for many applications high power systems, under a variety of operating conditions, pulsed, continuous wave, rep-rated or burst mode of varying duty cycles.

What is really power optics? In what way do they differ from ordinary optics widely used in cameras, motion-picture projectors, i.e., everyday use? In the late 60s, scientific groups discovered an undesirable consequence of the thermal deformation of optical elements and surfaces during their interaction with powerful incident laser radiation [1-5]. Undoubtedly, the physical effect observed should have limited many applications requiring high power optical systems under development at that time. The gist of the effect was the following: the optical surface of even a very perfect mirror does not fully reflect radiation that is incident on the surface. A small part of energy, about 0.5–1.0%, is absorbed by the mirror and is transformed into heat. With an increase of laser beam power, even a small absorption is enough to cause an undesirable thermomechanical stress in the mirror, and can distort the geometrical shape of the reflecting surface. Standard requirement for optical surface deformation is $< \lambda/20$, where λ – laser radiation wavelength. This, in turn, will influence, for example, the performance and optical characteristics of the laser system. We know that thermal deformations of mirrors can not only degrade the basic beam quality but can as well lead to the cessation of lasing in some cases. The response can be either transient or permanent, i.e., catastrophic.

Archaeologists have indicated that man made mirrors five thousand years ago. First, they were made of polished bronze or silver plates. In Roman times, there appeared glass mirrors with a tin or lead coating. Since that time, the technology of mirror production has changed many times. Most are either glass or quartz with applied metallic or dielectric coatings. Their early use was limited to low intensity sources, e.g., astronomical observations, illumination, etc.

The occurrence of mirror failure under intense laser radiation appeared from the very beginning with lasers. In lasers of all types, lasing occurs in a resonator that consists minimally of a pair of optical mirrors, and through one coherent radiation is emitted. In the early development stages these resonator cavity elements were nothing more than quartz substrates with a mirror coating. However, lately the power of the laser has increased hundreds, even thousands of times higher. Today the fabrication of mirrors capable of functioning while keeping their design characteristics under the influence of intense coherent radiation is one of the key challenges in the development of improved lasers of practical usage. The principal question is whether it is really possible to create mirrors that are not damaged by the powerful incident radiation flow when using the output for practical purposes, e.g., material processing. For contemporary laser systems this value in the resonator can be on the level of 200-300 kW/cm2. Perhaps it is appropriate to suggest a useful definition of damage here. We propose an applied or operative one, namely, damage is said to have occurred when the element or system no longer performs the function for which it was intended within specified limits (whether or not there is permanent or catastrophic damage).

Numerous institutions throughout the world have studied various aspects of these power optics questions from the beginning of the dawning of lasers. In fact, a Symposium on Optical Materials for High Power Lasers and GCL/HPL has been held annually in the United States and in Europe since 1969. Likewise a related conference "Nonresonant Laser Matter Interactions" also started in 1969 has addressed this subject in Russia. These meetings and their proceedings are the principal forum and repository for research and development activities. The material is readily available. Interest in the subject continues unabated to today. It is their goal, i.e., preventing lasers from committing suicide and demonstrating a high threshold for optical serviceability.

The real power density or the intensity of light at which distortions reach a preset limit can be a defined threshold of optical serviceability. One is not allowed to exceed this threshold because of elastic deformations of an optical element for the intensity distribution of the laser radiation will be altered unacceptably. If the intensity is further increased, deformations can be transferred into a non-elastic area and become plastic, and thus result in a non-recoverable effect in the element. The element is said to have failed. So, it is required to fabricate elements, which over a long period of time, can withstand high optical loads, e.g., from a few hundreds of watts up to several kilowatts per square centimeter of the surface for CW or P-P with high average power scenarios.

To imagine how difficult this task is, it is enough to describe two examples. Let us take the case of a mirror in an experimenter's hands for only a few seconds. Deformation due to variations in heating from the hand's warmth, of the optical surface can increase the limiting tolerated value. But in this case if the mirror is left to "cool," its original shape is generally fully recovered. In reality, mirrors of advanced technological lasers are routinely exposed to power densities of a few kilowatts per square centimeter. This level can be compared with the heat that is radiated into the surrounding space directly from a unit of the Sun's surface. Thus, it follows that if we "put" a laser mirror "on the Sun," the shape of its surface would not change by more than a micrometer. These types of mirrors are similar to those required for powerful industrial lasers.

To create these power optics, a number of problems related to quantum electronics, optics, thermoelasticity and heat exchange, materials study and advanced technologies, etc., need to be addressed. The first step in this might be to substitute a semitransparent quartz disk by a metallic one and to extract the radiation externally by diffraction through an aperture in the mirror or at its boundary. Metals exhibit nearly perfect reflectivity in the infrared. They also possess high thermal conductivity, which means they can transfer heat from the zone of incident interaction on the mirror surface. However, pure metals have some disadvantages as well, e.g., a high coefficient of thermal expansion—they can easily change their size or shape when heated—and also a low hardness to which it is generally difficult to polish a dielectric such as quartz. However, recent advances have allowed for excellent finishing of ceramics such as silicon carbide, which makes for quite excellent mirror substrates (providing quality coatings are applied). It should be noted that advancements in single point micromachining by diamond turning and optical polishing of metals are in many ways greatly alleviating general laser mirror problems [6-14]. It is interesting to recall an old occurrence (early 70`s) when physics was applied for the first time came to metallic optics. Consider a request to polish a metal disk in one and the same room as with quartz or Iceland spar. This was simply ridiculous as metallic dust in the workshop where final polishing was performed on precision optics, was not advisable. But nevertheless, a way out was found—a fully metallic mirror (Figure 1).

Figure 1. Metal based high quality optics.

Investigating metals and alloys of possible use for mass production, it was demonstrated that one could increase the threshold of optical serviceability of these new mirrors by ten times as compared to a traditional quartz element. The fact that its suitability was increased by "ten times" was not enough. It became obvious that the required level of power density and thermal loading of the mirror of a very powerful laser could be reached only with the assistance of substrate cooling.

During cooling by a circulating liquid (usually water, kerosene, alcohol, silicone, etc.), the heat withdrawn is directly proportional to the difference of temperatures of an inert body and the heat carrier. It might not be too difficult to remove thermal power rates at kilowatts per square centimeter if the mirror is heated up to a temperature of about one thousand degrees. However, it is hardly possible to realize a "perfect" optical quality for the mirror surface. There is a contradiction—convective heat exchange is greater at high temperatures, while temperatures close to ambient are desired for the stability of the mirror's geometrical shape (and other thermo-optical characteristics of the mirror). It is possible however to resolve this contradiction by means of a more efficient heat removal technique. This task was not demonstrated at temperatures of optical surfaces at about 100°C or less (typical for a laser mirror). Experiments were initiated by milling in the back side of a metal substrate channels through which the tap water was circulated. These cooling channels were placed as close to the surface as possible, but they frequently led to a vibration and deformation due to the water movement. One attempt to resolve this was to make the channels shallower. This approach led to the conclusion that it is much better to use porous capillary substrates that upon sectioning appear like a form rubber. Now the heat exchange was more intensive both due to a large surface thermal cooling as well as an increased mixing of the cooling fluid that travels in the microcapillaries. Moreover, the matrix–skeleton of the porous body functioned as a closely spaced mirror surface supporting structure, allowing for improved stability in the mirror surface, i.e., it maintains its initial geometry better (Figure 2). In this situation, the cooling fluid engenders a chaotic river flow. The matrix of the porous body resembles the structure of "lace" retaining the smooth character of the surface by the large number of closely spaced supports [15-17]. Additionally one usually also applied a high reflectance coating on the highly porous heat exchanger substrate to further reduce the thermal load on the cooled substrate. Then the coating is polished if necessary to achieve the desired reflectivity and figure. The thickness of such coatings should be very small—100-150mkm; otherwise, they can reduce the heat absorbed by the mirror surface. There are several methods used in applying quality multilayers on these highly porous cooled metal substrates. It was frequently solved by employing intermetallics, i.e., chemical compounds of metals. These intermetallic coatings can be achieved, for example, by precipitation from a gas phase (channel vapor deposition, CVD). In such a way it is possible not only to realize a fine separation layer, but if necessary, to restore mirror quality. Intermetallic coatings have another important property: their structure makes it possible to obtain mirror surfaces of very high optical quality [18]. If one inspects a polished ordinary metal through a microscope, its surface frequently reminds one of an orange peel, i.e., it is covered with hillocks and cavities. To reduce this micro-structure, additional finishing measures can be employed, e.g., before final polishing the metal substrate can be alloyed to make it not only harder, but more fine grained. Nevertheless, the final scale of unevenness remains too large, i.e., from 0.01 to 0.1mkm. The structure of intermetallic coatings from the start is very fine, about 0.1mkm and after treatment by a diamond tool or by optical polishing, it is possible to realize almost perfect mirror surfaces—with an average scale size of only a few thousandth of a micrometer.

Thus, a solution was found. Mirrors for powerful "industrial" lasers of continuous and or pulsed formats can employ mirrors containing highly porous heat exchangers, together with the use of a fine separating layer and with a reflecting coating. Industrial lasers with such mirrors have been used successfully for welding, cutting and the hardening of metals (Figure 3).

There are still possibilities to further improve these "power optics." By changing the pressure of the coolant, it is possible to force it to boil at room temperature. This is within the conditions similar to those of the final treatment of the mirror's surface. During the boiling of the coolant, part of the absorbed heat is used for steam formation, thus heat removal efficiency is tens to hundreds of times higher than within convective heat transfer. Steam can easily penetrate into porous capillary substrate structure.

Figure 2. First high power/energy water cooled porous mirrors.

Figure 3. Laser cut examples of technology (glass, composite, stone, sitall).

Additionally, water can be substituted by a liquid metal. For example, alloys of sodium, potassium and cesium, which have low melting temperatures. The efficiency of the heat removal will increase, since the heat would not only be carried away by the moving liquid, but by being transferred to the heated metal itself as well, which would be an almost perfect heat conductor, as metals usually are. Liquid metal heat conductivity allows the removal of heat from reflecting surfaces at a few kilowatts per square centimeter on the surface.

Eventually of course as one approaches "perfect" reflectivity for coatings—cooling could be eliminated entirely except to stabilize or homogenize the temperature.

It is interesting to ask if there is a limit at present for the amount of heat that can be withdrawn from the surface into the mirror substrates. From the point of view of present day power laser optics, the limited heat regime is determined by gas breakdown at the mirror surface, i.e., by plasma formation in the resonator cavity. The developed and demonstrated methods of intensive heat removal may in turn be useful in other fields where there are no such limitations. Using the methods of metallic mirror cooling it is possible to solve, for example, "non-mirror" specific problems, like the cooling of large integrated structures for the present day anodes of powerful X-ray lithographic installations for microelectronics manufacturers. These devices should endure heat loads of between a few tens and hundreds of kilowatts per square centimeter. The highest value of evacuated heat flax, realized in the case of high power optics was measured on the level of 8,2 kW/cm2 (82 MW/m2). An important issue in the generation of power optics can be economical by implementation of a number of technical and beneficial solutions for the national economy [19-22]. Among them, for example, can be diamond turning and the concurrent creation of a large range of machines for high pressure, accuracy and uniformity of treatment. With the help, e.g., of machines that make memory disks, as well as drums for copy machines and other high tolerance equipment. For "non-mirror" specific uses, an obvious application is related to the development of technologies using these various capillary-porous structures plus the application of environmentally benign coatings of high hardness on metals and intermetallics.

Consider the integration of a big number of laser diodes (LD) into 1-d and 2-d structures radiating both non-coherent and coherent laser radiation of high power. The difficulty of such integration of LD is the necessity to retain the temperature of radiating hetero-junctions of LD within a narrow temperature range, insuring in turn the frequency stability of the laser radiation. Heretofore, values of thermal power density at the level of few kilowatts per square centimeter of heat exchange surface at room temperatures were required. In practice, such array structures of LD consist of a large number of LD soldered to the surface of a perfectly prepared metal mirror at a high packing density on the above mentioned radiating elements. As the array radiates high intensity laser light (even at today's demonstrated efficiency greater than 60%), heat exchangers should extract, as indicated above, heat flows from an active medium $Q > 1000$ W/cm2. At this level, the displacement of the radiation spectrum conditioned by a thermal increase of the radiating layer should not increase more than 3 nm relative to the initial wavelength of lasing, corresponding to a change of temperature of the active layer not more than by 10°C. This is why the heat exchanger of such a device should have a relatively low thermal resistance of not more than 0.1 K/W.

To obtain these high values of heat removal in the devices, high thermal conduction materials, such as: beryllium ceramics (BeO), (K= 3.7 W/cm°C), diamond (K= 20 W/cm°C) etc. should be employed. Unfortunately, the effort and cost to produce and treat these materials makes the process of array fabrication more difficult and expensive. Silicon carbide (SiC) is frequently used as a high thermal conductivity material. Besides high thermal conductivity (in the best cases it is close to copper's thermal conductivity) silicon carbide has enough electrical impedance, it can be best treated and is safe from the point of the environment, it also has a high hardness that is important during optical polishing. Both separate and combined heat removal elements can be made from silicon carbide as well as complete microchanneled or porous heat exchangers. Use of silicon carbide as a thermal heat sink material is also convenient because its coefficient of thermal expansion is close to that of GaAs—the basis of numerous laser diode compositions. This aids to prevent the material from cracking during, e.g., soldering. It should be noted that development of optical grade silicon carbide requisite size is available for solving some optics applications in high power laser systems and for astronomical purposes (Figure 4).

Figure 4. Big size SiC mirror under polishing.

The results obtained during the last decade at Russia and the United States are fully consistent with the current trends in the development of the market for LD technologies [23-25].

Indeed, as early as 1991 much attention has been concentrated on the following trends in the development of technologies, which are so popular nowadays in research groups and industrial companies throughout the world:

- development of efficient heat-sinking systems for 1-d and 2-d LD arrays;
- improvement of the soldering technology for a linear array and a heat-sink component taking into account the problem of minimizing the thermal resistance and of bending of the 1-d array itself;
- creation of a new laser system based on a phase locked 1-d array of LD;
- use of phase-locked systems with efficient injection of radiation into a fiber;
- phase locking of 2-d LD arrays (not realized yet);
- efficient beam steering for the case of high power 2-d matrix of LD radiation (not realized yet);
- development of new configuration of high power solid state lasers – single module scalable disk laser with diameter >10cm (not realized yet);

The above trends require maximum concentration of intellectual and financial resources. We already understand the physical and technical problems encountered in the construction of high heat-sink components. The evident advantages of technologies involving phase-locking of LD arrays have been demonstrated, which has made it possible to develop new approaches and to implement a range of ideas included in the above list.

CONCLUSION

One very important and true to life relationship should be mentioned in this rather difficult time for science. The resources invested effectively in the development of any field of advanced technology will as a rule afford a feedback in a number of ancillary applications in other ancillary and sometimes rather remote fields of science and technology. Thus, phase-locked 1-d and 2-d arrays of LD with high levels of radiation and new configuration of solid-state laser – single module scalable disk lasers – appear, along with many other innovations mentioned at the Symposium HPLS@A-2012, September 10-14, due in part to achievements in the field of high power optics [26]. Power optics is a universally recognized contributor to other advanced laser systems and applications for the 21st century. Russia and the United States are leaders in this field of high power optics.

REFERENCES

[1] V. V. Apollonov, A. I. Barchukov, A. M. Prokhorov, High power CO_2-laser, Report of FIAN,1970.

[2] V. V. Apollonov, A. I. Barchukov, A. M. Prokhorov, Thermal deformation of solid state surface by laser beam, *Letters to JETP,* т. 15, 1972.

[3] V. V. Apollonov, A. I. Barchukov, A. M. Prokhorov, Optical mirrors loses on scattering measurements, QE, т. 4, № 16, 1973.

[4] V. V. Apollonov, A. I. Barchukov, A. M. Prokhorov, *CO_2 laser mirrors investigations,* Report of FIAN, 1973.

[5] V. V. Apollonov, A. I. Barchukov, A. M. Prokhorov, Thermal deformation of laser mirrors, Second European conf. “Lasers and applications”, Drezden, July, 1973.

[6] V. V. Apollonov, A. I. Barchukov, A. M. Prokhorov, Optical characteristics of laser mirrors, Preprint № 157, FIAN, 1973.

[7] V. V. Apollonov, A. I. Barchukov, A. M. Prokhorov, Thermal deformation of laser mirrors, *Radiotechnics and electronics*, V. XIX, № 1, 1974.

[8] V. V. Apollonov, A. I. Barchukov, A. M. Prokhorov Thermal action of laser beam on the surface of laser mirror, Preprint № 105, FIAN, 1974.

[9] V. V. Apollonov, A. I. Barchukov, A. M. Prokhorov, Optical destruction of heated laser mirrors in CO2-laser systems, *JEEE of QE*, №6, 10, 1974, p. 505.

[10] V. V. Apollonov, A. I. Barchukov, A. M. Prokhorov, *Thermal action of high power laser on the surface of solids*, QE, т. 2, № 2, 1975.

[11] V. V. Apollonov, A. I. Barchukov, A. M. Prokhorov, Anastigmat I, Report of FIAN, 1974.

[12] V. V. Apollonov, A. I. Barchukov, A. M. Prokhorov, *Absorption coefficients measurements* at λ=10,6 мкм, Preprint № 52, FIAN, 1975.

[13] V. V. Apollonov, A. I. Barchukov, V. K. Konykhov, A. M. Prokhorov, Figures of merit for mirrors, *Letters to JTP,* v. 1, p. 11, 1975.

[14] V. V. Apollonov, A. I. Barchukov, A. M. Prokhorov, R Anastigmat II, Report of FIAN, 1975.

[15] V. V. Apollonov, A. I. Barchukov, A. M. Prokhorov, High intensity porous metal mirror, Report of FIAN, 1977.

[16] V. V. Apollonov, A. M. Prokhorov, Vapourtron effect for laser mirrors, *Letters to JTP,* v. 4, p. 8, 1978.

[17] V. V. Apollonov, A. I. Barchukov, A. M. Prokhorov, Porous structures for high power optics, *Letters to JTP,* v. 4, p. 19, 1978.

[18] V. V. Apollonov, A. I. Barchukov, V. K. Konykhov, A. M. Prokhorov, *Intermetallic coatings for laser mirrors,* QE, т. 5, № 2, 1978.

[19] V. V. Apollonov, A. I. Barchukov, V. K. Konykhov, A. M. Prokhorov Beam divider for high power laser beam, Preprint № 136, 1978.

[20] V. V. Apollonov, F. V. Bunkin, S. A. Chetkin, V. Yu. Khomich, Thermodeformation as a method of laser beam energy distribution measurements, Letters to JTP, v. 4, p. 17, 1978.

[21] V. V. Apollonov, P. I. Bystrov, A. G. Goncharov, A. M. Prokhorov Liquid metals as a heat carriers for laser optics. QE, № 3, 1981.

[22] V. V. Apollonov A. M. Prokhorov et al, New heat carriers for high power optics, Letters to JTP, 1981.

[23] V. V. Apollonov, A. M. Prokhorov, A. H. Guenther, Power Optics, Problems, Developments and Opportunities, *Laser Physics* V. 11, N.8, 2001.

[24] V. V. Apollonov, A. M. Prokhorov, A. H. Guenther, Mirrors maintain shape under intense laser light, Laser Focus World January, p. 101, 2003.

[25] V. V. Apollonov, *High power optics and it`s new appearances,* Oboronzakaz, March 2012, p.47.

[26] V. V. Apollonov, High power optics, Symposium HPLS@A-2012, Istanbul 10-14 September..

Chapter 15

NEW MATERIAL SIC-NEW TECHNOLOGIES

HIGH-POWER SEMICONDUCTOR LD STRUCTURES AND NEW APPLICATION

At present, high-power/energy lasers are often treated as laboratory equipment requiring constant maintenance by specially trained personnel. In this regard, the most promising now are semiconductor lasers. To date, they have record-high efficiency reaching 70%. Modern technologies of fabrication of semiconductor structures have made it possible to significantly increase the lifetime of the laser, which is tens of thousands of hours of continuous operation. Besides, the use of a different semiconductor material can change the wavelength range of the radiation from the infrared to the ultraviolet.

Currently, semiconductor lasers are the only lasers used in household appliances, namely, in communication and electronic devices. The cost of semiconductor lasers is still high, but decreases from year to year as the production rate grows. Semiconductor lasers are very compact and easy to operate.

The main feature of lasers based on semiconductor structures (LD-structure), which allowed them to work at room temperature, is a small amount of the active medium. Therefore, the output power of a single diode is low. Currently, it does not exceed 10 W. To increase the output power, one needs to use simultaneously a large number of laser diodes, which form one-dimensional (bars) or two-dimensional (arrays) structures. This direction in the development of high-power semiconductor laser systems is being exploited relatively rapidly nowadays.

An assembly of laser diodes has almost all the remarkable properties of single semiconductor lasers, such as high intensity, high reliability, and long lifetime (up to 10^{12} cycles). These lasers have smaller weight and size dimensions (up to 10 times) in comparison with other types of lasers. They can be easily fabricated with an independent low-voltage power supply (solar, nuclear) without bulky transformers. Equipment based on laser diode assemblies really becomes a reliable high-performance tool that can be used in industry, medicine, and military hardware.

Almost all developed countries are, in one way or another, involved in the design and development of such laser systems. The leading role is played by the USA, Germany, England, and France. Interest in the creation of these systems is huge. Suffice it to say that just in the last two years, many small companies, specializing exclusively in the production of these lasers, have appeared in the U.S. This proves that their commercial benefits come to the fore.

Thus, we can now single out three main areas of use of high-power laser diode assemblies.

Major Trends in the Development of Pulse-Periodic Lasers

Pumping of Solid-State Crystal Lasers

The use of LD-structures for pumping solid-state lasers leads to high output optical. The efficiency of LD-structures is low due to the low efficiency of solid-state lasers; however, pumping by semiconductor lasers makes it maximal. Radiation of semiconductor lasers does not deteriorate the working medium of crystal lasers, which makes it possible to significantly extend the lifetime of the entire laser system. Weight and dimensions of such systems are small, which allows them to be easily transported. The main purpose is to develop and fabricate diode arrays (bars) with the highest possible power density. Record-high values, i.e., 100 – 115 W from a 1-cm-long single diode array, were obtained by Coherent Inc., USA. The limiting power from one bar, as calculations show, can reach 500 W. At the Department of High-Power Lasers of the A.M. Prokhorov General Physics Institute of the Russian Academy of Sciences, a 1-cm-long diode array with an 80-W output was fabricated. Solid-state diode-pumped lasers are being increasingly studied and developed by a large number of firms in the various countries. High-power diode-pumped laser systems are being developed mainly in such major companies as Lawrence Livermore National Laboratory (USA), Rofin-Sinar Technologies Inc. (Germany), Coherent (USA), and Newport Corporation (USA).

Coupling Radiation from a Diode Array into Fiber

An important problem here is the coupling of radiation from a diode array into fiber for further transport to a work piece to be processed. Radiation losses in such a laser – fiber system reach 20% – 30%. At the same time, 200 ÷ 300-W laser systems are known to be produced industrially in the U.S. and Germany.

Phase-Locking of a Diode Array

In our opinion, the most promising application of laser diode assemblies is the creation of a phase-locked diode array. The output radiation of such a laser is a set of narrowly focused interference peaks (in particular one peak) with small divergence. The estimated output power can reach the power of a conventional diode array producing one set of beams that are out of phase with those of the other set of beams. This laser is a light source offering high efficiency and low divergence that are characteristic of semiconductor lasers and solid-state crystal lasers, respectively. By analogy with the technique of phase-locked antennas in the radio range, the beam divergence in the case of phase-locked diode arrays can easily be increased to 10^{-6} rad without expensive telescopic systems, which provides ultra-long communication and transmission of energy over long distances.

At present, these systems are being actively studied. For example, Energomashtekhnika Ltd. has been conducting research in this field since 1997 and has produced a prototype of a ~20W phase-locked laser. The possibility has been shown of scaling the process of phase-locking of the arrays and bars of laser diodes. Arrays of laser diodes with an output power up to 500 W have been fabricated. The effect of phase-locking of high-power laser diode structures is being increasingly investigated. According to the press in other countries, the maximum output power in a phase-locked array is reported to be only 1.5 W.

LASER SYSTEMS OF PROPELLANT IGNITION

One of the most important applications of laser diodes is the creation of a system of multiple ignition of rocket propellant in the combustion chamber of a launcher. Apart from the fact that the parameters of the laser system should provide multiple ignition of the mixture with 100% certainty, the laser design must meet the requirements specific to onboard lasers that are currently used in space technology:

1. Compactness.
2. Small size and weight.
3. Absence of high-voltage in the power supply system.
4. Resistance to vibrations and g forces.
5. Absence of active cooling (including the closed circuit with a refrigerant and a pump).
6. Long service life.

A new and specific requirement for the launcher-related design of the laser is the need to deliver the laser radiation into the combustion chamber through an optical fiber. Accordingly, the peak and average power of the laser must permit delivering its radiation into the fiber with a minimal chance of fiber damage.

Since July 2002, works have been on the way to create a laser system ensuring multiple ignitions of rocket propellant components in the prechamber (or ignitor) of the third-stage rocket engine. In the course of these works, the following sources of laser radiation were used:

(1) a solid-state laser operating in a modified spike regime, first proposed and patented by Energomashtekhnika Ltd., with the following radiation parameters:
wavelength, 1.054 μm;
output energy, 4 – 10 J;
temporal shape of radiation, a train of 100 – 150 pulses with an irregular period;
pulse train repetition rate, a single pulse;
duration of the pulse train, 600 – 700 ms;
duration of a single pulse, 200 – 400 ns;
pulse spacing, 4 – 7 ms;

peak output power, ~ 100 kW;
divergence, 4×10^{-4} rad; and
type of the active element pumping, lamp pumping.

Figure 1 shows the oscillograms of a train of pulses (Figure 1a) and of single spike (Figure 1b) of a laser operating in modified spike regime. A characteristic feature of the temporal shape is the irregular spacing between neighboring spikes within the pulse train and the high contrast of the spikes with respect to the pedestal (significantly higher than in the pulse of a conventional free-running solid-state laser). It should be noted that the duration of the spikes is about an order of magnitude shorter than in the traditional pulse. These differences in the temporal shape of the modified spike regime made it possible to realize a much higher peak output power compared to a traditional pulse of the same energy.

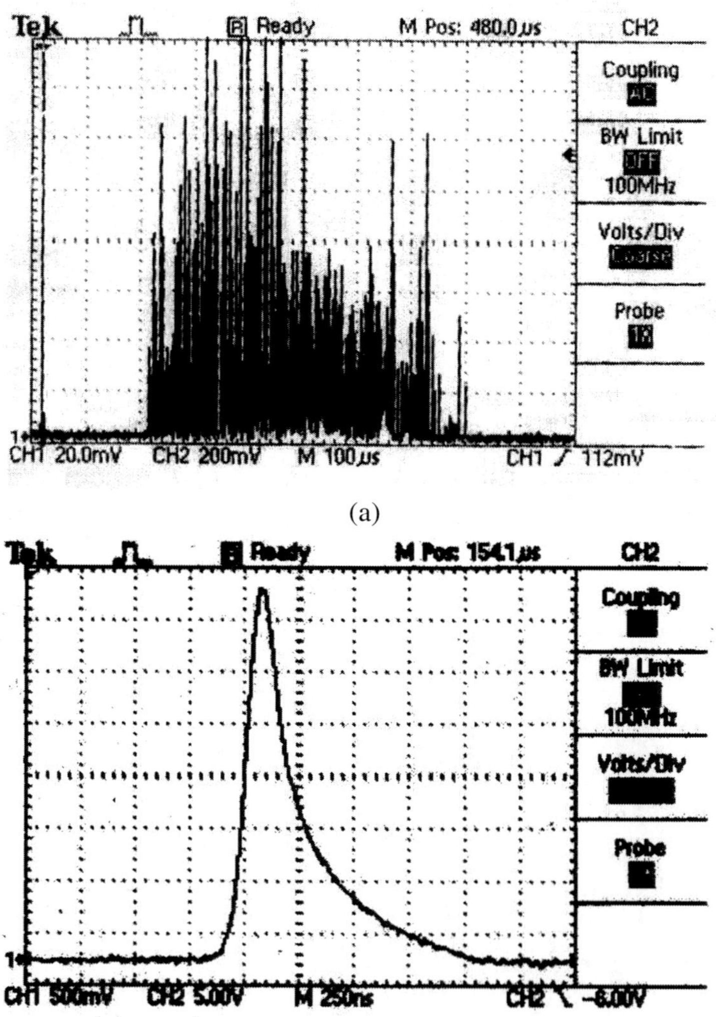

(a)

(b)

Figure 1. Oscillograms of (a) a "ruff pulse" (100 µs/div) and (b) a single pulse from "ruff pulse" structure (250 ns/div).

The laser is a laboratory model with a rather bulky power supply and cooling module. The dimensions of the laser head are as follows: 200 × 200 × 1400 mm. The appearance of the laser head is shown in Figure 2.

(2) A solid-state *Q*-switched pulsed laser:
wavelength, 1.06 μm;
pulse energy, up to 1 J;
peak power, up to 100 MW;
pulse duration, 10 ns; and
repetition rate, a single pulse or 10 Hz.
The laser however was a rather cumbersome laboratory bench.

(3) A serial ytterbium fiber laser (YPL1/100/20):
wavelength, 1.062 μm;
average output power, 20 W;
pulse duration, 120 – 150 ns;
pulse repetition rate, 20 – 80 kHz;
maximum pulse energy, 1 mJ;
maximum peak power, 10 kW; and
divergence, $10^{-3} \div 3\times10^{-3}$ rad

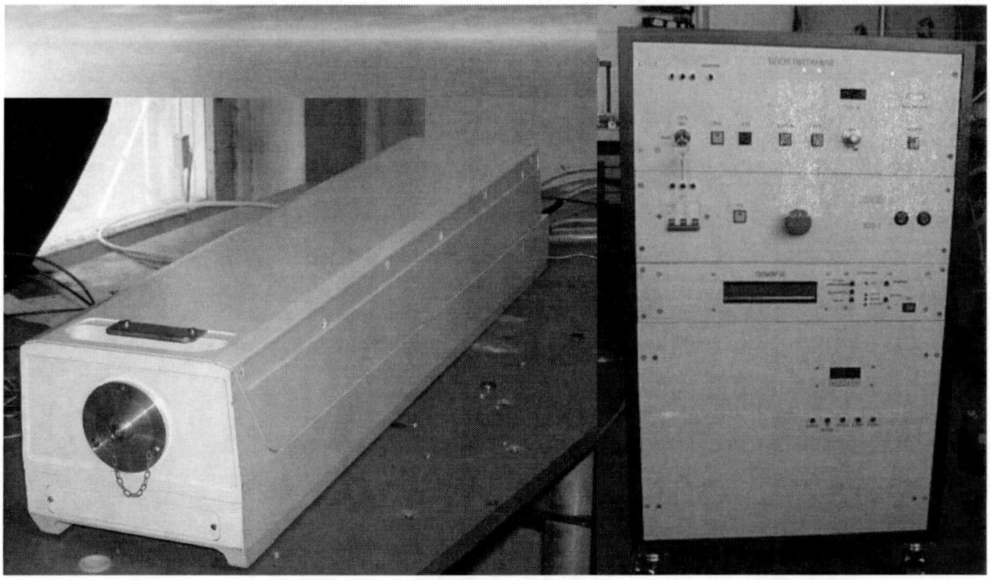

Figure 2. Laser head and power supply.

The laser was significantly smaller in size than the two above-mentioned lasers, the total weight of the power supply unit and the laser head being 8 kg. Neither water cooling nor a high-voltage power supply was used in the laser.

Results of the Experiments

Experiments were performed to study the ignition of different fuel mixtures (oxygen – kerosene, oxygen – alcohol, oxygen – hydrogen) in the prechamber (or ignitor) of the third-stage rocket engine. In the first stage of the research, the radiation was delivered into the prechamber through an optical window, which was installed in the seat of a standard spark plug. The radiation was focused by a lens mounted outside the chamber onto the target placed inside the chamber. In the experiments a metal mesh was installed in the channel of the input laser beam. This target made it possible to successfully ignite the fuel mixture. Due to the large capillary resistance, the target in the form of a mesh effectively protected the sapphire window (through which the laser beam is delivered) from condensation. The disadvantages of this target include the problem of maintaining its thermal stability during the operation of the combustion chamber. A graphite rod, mounted in a brass ring, is inserted into the combustion chamber. This target can be moved along the axis of the combustion chamber, thereby determining the optimal zones to initiate ignition. This target makes it possible to ignite the fuel mixture at very low levels of laser energy. The downside of the target is the possibility of its destruction in a highly oxidizing environment; the wall of the subsonic part of the exit nozzle is covered with oxides, and special targets in the combustion chamber are absent. The laser beam was focused on the wall with a long-focus lens. This variant is the most reliable one for ignition, because no high requirements are imposed on focusing and there are no changes in the scheme of the combustion chamber.

The experimental results showed a significant advantage of laser (1) over other lasers: the use of this laser made it possible to ignite all the above-mentioned mixtures with a large margin of the radiation energy. In particular, the ignition threshold for the oxygen – hydrogen mixture was 90 mJ.

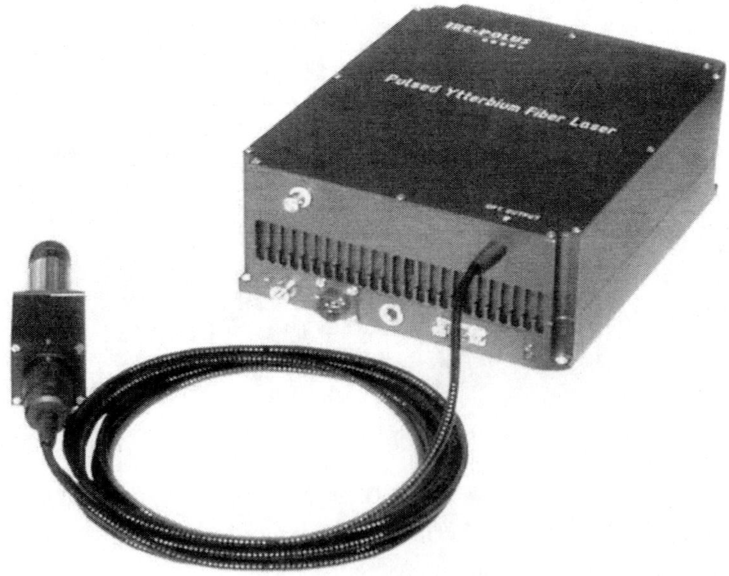

Figure 3. YPL1/100/20 laser head.

Laser (2) ignited only the hydrogen – oxygen mixture (this mixture is known to be the most flammable one), and the ignition threshold was 60 mJ.

Laser (3) ignited all the above-mentioned mixtures. However, reliable ignition was achieved using a total laser power of 20 W. Moreover, for the ignition to be successful the optimal repetition rate should be equal to 50 kHz and the target should be of conical shape to provide accumulation of the material evaporated from the target and, accordingly, its additional heating.

In our opinion, in experiments with two lasers (1) and (3), the optical breakdown threshold for the gas mixture near the target was not reached; the basic mechanism of ignition was the contact of the gas mixture with heated vapors of the target material formed near it by laser ablation. Because in the case of laser (1), the peak power was about 10 times higher, the higher was the temperature of the vaporized material. Most likely, this parameter played a crucial role in the successful ignition of less flammable mixtures such as oxygen – kerosene and oxygen – alcohol.

Because of a significantly higher peak power, in experiments with laser (2) the optical breakdown threshold of the gas mixture was substantially exceeded. The plasma, produced near the target on the leading edge of the laser pulse, shields it from the rest of the pulse and no target material is significantly evaporated. Therefore, the main ignition mechanism is the ignition of the gas mixture at its contact with the laser plasma. However, based on the experimental results, this type of ignition is less effective because it does not allow one to ignite less flammable mixtures. Based on these experiments, we can draw the following conclusions:

1. The mechanism of laser ablation is more effective to ignite the rocket propellant mixtures than the optical breakdown of the mixture. An additional argument in favor of ablation is the fact that the propellant itself is not involved in the process. Therefore, hot vapors of the target can be produced either before or during the injection of the propellant components. In the case of the optical breakdown of the mixture, its threshold depends strongly on the gas composition and pressure in the chamber. In this case, the breakdown can occur after filling the chamber with the mixture. This regime is ideal in the case of a small combustion prechamber, but can lead to self-ignition of the mixture and to an explosion in a large combustion chamber.
2. The laser pulse energy (and, hence, its peak power) significantly affects the temperature of the evaporated material, and the total energy of the pulse train influences its mass. An increase both in temperature and mass of the evaporated material provides more reliable ignition of propellant mixtures. Therefore, in designing an ideal laser for ignition, one needs to optimize both the peak power of a single pulse and the energy of the pulse train (in the case of a pulse-periodic laser, high power).
3. It is also likely that an important parameter is the laser pulse repetition rate. At some optimal repetition rate, additional heating of the target vapors by subsequent pulses is also possible, but in such a way that the target vapors would not fully absorb the incident radiation and allow for ablation in all the pulses of the train. Thus, when working with laser (3), optimal was the frequency of 50 kHz, at which a continuous optical discharge was observed near the target (Figure 4).

Figure 4. Photo of an optical igniter with optical discharge.

Fiber-Optic Coupling of Radiation and Its Advantages

The use of optical fibers for coupling light into the combustion chamber eliminates the need to install an optical window in the wall of the chamber. Moreover, optical fibers deliver radiation through a small hole in the housing, comparable in diameter to that used for the injection of fuel mixtures of components (hundreds of μm). Installation of an optical window is often impossible due to design reasons; besides, an additional risk is associated with the possibility of explosion of the window at very high temperatures and pressures in the combustion chamber.

Therefore, in the second stage of the works, an original construction has been designed and manufactured to couple optical radiation into fiber and place fiber inside the combustion chamber. This design ensures high reliability of the fiber attachment, long service life without re-polishing of the fiber end located in the combustion chamber, and the optimal delivery of radiation to the target. The optical scheme of the ignition system with fiber-optic input is shown in Figures 5–6. Subsequently, this design was used to deliver the radiation of lasers (1) and (3). The radiation of laser (2) could not be coupled into the fiber due to a very high peak power that damages the fiber end. The power can be split into 7 channels and delivered to the igniter.

Thus, a single-pulse *Q*-switched solid-state laser with a pulse duration of ~10 ns is not optimal for igniting propellant mixtures and causes big problems when coupling the laser beam into an optical fiber. We consider it inappropriate to use such lasers for igniting propellants. Based on the above considerations, it is proposed to test sequentially the lasers developed and designed by Energomashtekhnika Ltd. for igniting propellants in the combustion chamber of the rocket booster (the oxygen – hydrogen mixture). The lasers have the following parameters:

I. A solid-state laser with semiconductor pumping:
wavelength, 1.06 μm;
pulse duration, 100 – 200 ns;
pulse energy, 10 – 50 mJ;
peak power, 50 – 500 kW
pulse repetition rate, 5 – 100 μs (the optimal interval is determined for each working mixture); and
continuous operation time, 5 s.

Based on the experience in designing solid-state diode-pumped lasers and laser diode arrays at Energomashtekhnika Ltd., we came to the conclusion that such parameters are offered by a transversely diode-pumped solid-state laser with a passive optical switch (e.g., LiF with color centers) to ensure modulation and by laser diode arrays in phase-locked and out-of-phase regimes. The proposed laser can make use of the original patented diode-pumped laser with the following parameters: pulse energy, 50 mJ; pulse repetition rate, 100 Hz; and a special design for fixing the active element and diode laser arrays, which eliminates water cooling. At present, the laser emits pulses with energy of 50 mJ and a repetition rate of 50 Hz. If long-term continuous operation is required (the laser operates from 1 to 5 s), a substantial increase in the pulse repetition rate (up to 1 kHz) is possible without reducing the pulse energy.

II. A laser diode array in out-of-phase oscillation mode (intermediate stage):
wavelength, 0.808 μm;
operation mode, pulse-periodic;
pulse duration, 100 μs;
pulse energy, 0.2 J;
peak power, 2 kW;
maximum peak power density during focusing, 2 MW/cm^2;
pulse repetition rate, 1 kHz; and
continuous operation time (without cooling), 5 s.
III. An array of laser diodes in phase-locked oscillation mode
wavelength, 0.940 μm;
operation mode, continuous;
power, 40 W;
maximum peak power density during focusing, 40 MW/cm^2; and
continuous operation time (without cooling), 5 s.

In comparison with diode-pumped solid-state lasers, the laser diode array, operating both in out-of-phase (cheaper) and phase-locked oscillation modes is even a more compact and reliable ignition device. Since, as shown above, the mechanism of laser ablation is more effective to ignite rocket propellants than the optical breakdown, the devices based on the array technology must take a lead in the solution of the above problems. It should be noted that the first samples of laser diode arrays developed for other purposes have already been produced at Energomashtekhnika Ltd. Currently, work is underway to ensure phase-locking of laser diode structures.

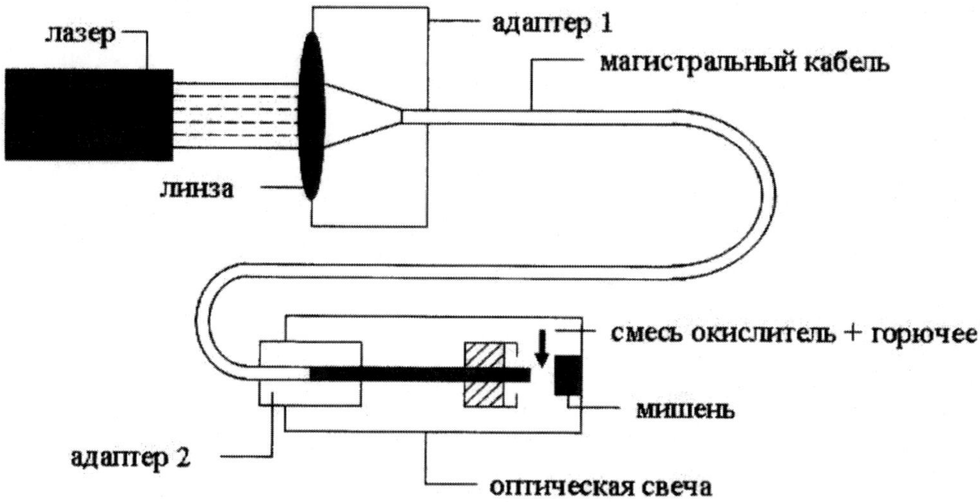

Figure 5. Fiber system for delivery of optical radiation to an optical igniter.

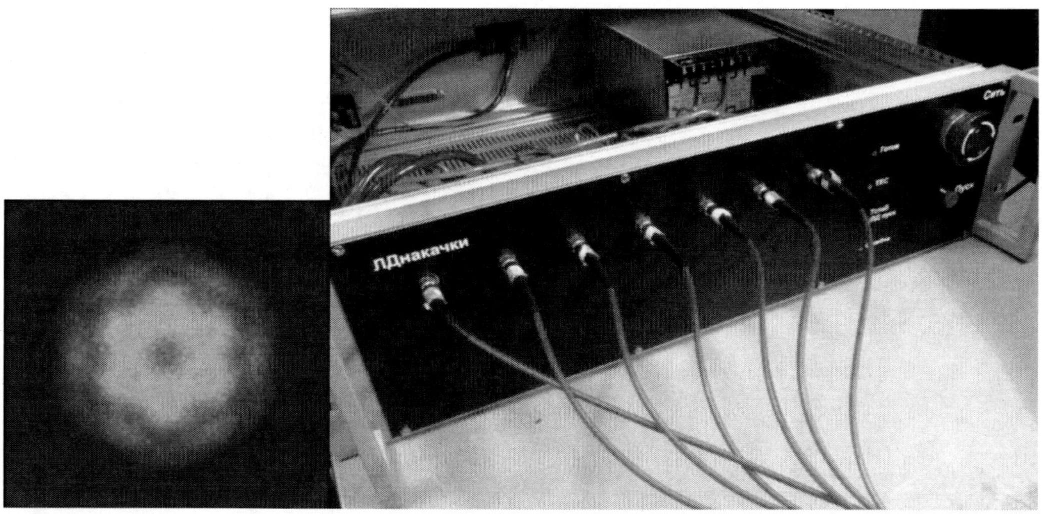

Figure 6. Seven-channel pumping block with a near-field pattern.

Estimates show that the radiation of the proposed lasers can be coupled into an optical fiber without the risk of damage. A significant increase in the pulse energy and pulse duration at a high repetition rate within 5 s or a transition to the continuous regime with tight focusing of the radiation under conditions of its phase-locking suggests that the proposed ultra-compact laser systems are ideal devices for solving the problem of reliable multiple ignition of various components of the rocket propellant in combustion chambers of the launchers.

Conclusion

Thus, high-power diode arrays are promising sources of radiation, the dominant role of which, in the new millennium, is ensured by a unique set of basic features and characteristics.

New SiC-Mirror Manufacturing Technology Capabilities

For many years, Russia has been developing new materials based on refractory metals and ceramics. Results of works have been widely used in the development of a nuclear jet engine and fuel elements of nuclear reactors. In the early 1970s, researchers at the Lebedev Physical Institute of the USSR Academy of Sciences (the General Physics Institute of the USSR Academy of Sciences was organized later in 1983 on the basis of the Laboratory of Oscillations of the Lebedev Physical Institute) discovered the physical effect of local deformation of the reflective mirror surface and began to actively conduct research to design power optics for high-power laser systems. The solution of this problem required a material that, under conditions of radiation exposure, alternating temperatures and inertial accelerations, retains high geometric stability and has high corrosion resistance in aggressive environments, such as hydrogen fluoride.

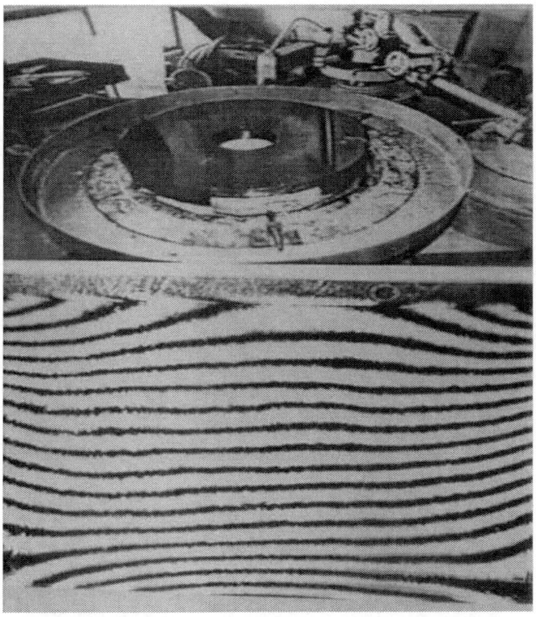

(a)

(b)

Figure 1. (a) SiC laser mirror (1 m in diameter) cooled via a porous heat exchanger and (b) reaction-bonded non-cooled SiC mirror.

Analysis of a wide range of materials on the basis of stability parameters proposed in [1, 2] has shown that silicon carbide, produced either by reaction bonding or sublimation has an advantage over other materials. Specific heat loads under CW, repetitively pulsed and pulsed irradiation also play an important role. The data obtained are in good agreement with the results of similar studies carried out much later by U.S. specialists [4, 5].

The technology we developed in the late 1980s allows one to fabricate mirrors with a high specific radiation resistance while maintaining the geometric stability of the reflecting surface (Figures 1a, b).

Table 1. Properties of different materials, and results of their evaluation by the stability criteria [1, 2]

Material	Properties				Constant loading Criterion (τ > 0.1)					Pulsed loading Criterion (τ < 0.1 s)			Sti ffn ess
	$\alpha \cdot 10^6$ K^{-1}	λ W/μK	C J/kg·K	$\gamma \cdot 10^{-3}$ kg/m³	$E \cdot 10^{-4}$ MPa	$\sigma_{0,01}$ MPa	$\lambda/\alpha E \cdot 10^{-6}$ W/m	$\frac{\sigma_{0,01}\lambda}{\alpha E} \cdot 10^{-4}$ W/m	q kW/cm²	$C \cdot \gamma/\alpha \cdot 10^{-10}$ W·s/m³	$\frac{\sigma_{0,01}(c\gamma\lambda)^{1/2}}{\alpha E} \cdot 10^{-5}$ W·s$^{1/2}$/m²	q kW/cm²	
1	2	3	4	5	6	7	8	9	10	11	12	13	14
Cu	17	400	380	8,9	13	185	24	3,4	2,6	20	31	4	14,7
Mo	5,1	130	260	10,2	32	280	25	2,2	1,6	52	34	5	32,3
W	4,6	170	134	19,3	41	750	37	6,8	2,5	56	83	10	21,2
SiC_{pc}	2,8	130	1200	2,7	40	230	46	2,7	2,3	116	43	5	131
SiC_c	2,8	230	1200	3,2	45	330	82	6,0	3	137	76	11	143

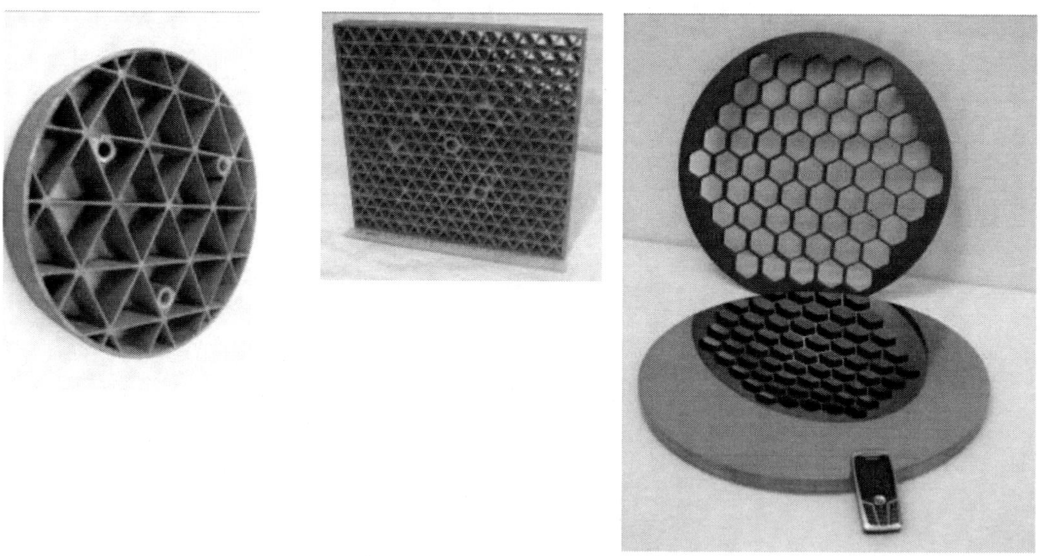

Figure 2. Honeycomb structure of lightweight laser mirrors.

Currently, SiC is recognized worldwide as a highly promising material for mirror substrate fabrication and development of all the telescope components due to its many attractive properties, such as high specific stiffness (high rigidity and low density); low thermal sensitivity (low thermal expansion coefficient and very high thermal conductivity);

and extreme hardness and stiffness, which allow for good mechanical property retention in the micron range and ultimate characteristics for mirror forming and polishing.

Currently available equipment makes it possible to produce meter-class monolithic silicon carbide mirror blanks. The methods developed for the manufacture of cellular silicon carbide substrates allow for fabrication of lightweight space- and ground-based laser mirrors (Figure 2).

The original welding technology of SiC ceramics enables the manufacture of complex structures having low weight and high rigidity, such as, for example, optical benches, mounts (racks, farms, etc.) and other elements of laser mirrors (Figure 3).

The developed welding technology provides a vacuum-tight connection, which enables the production of hermetic containers made of silicon carbide for long-term storage of spent nuclear fuel (SNF). Widely known is the inertness of silicon carbide to corrosion and high resistance to all types of ionizing radiation (gamma rays, neutrons, protons), which ensures reliable storage of SNF that is in contact with groundwater for thousands of years. In cooperation with the OJSC SPII VNIPIET we have designed and manufactured a full-size prototype of a silicon carbide cask for long-term storage and disposal of absorbing elements of nuclear submarine reactors and unreprocessible SNF of nuclear-powered ice-breaker fleet.

The methods of silicon carbide purification and high production standards elaborated in Russia made it possible to design and build tooling for manufacturing processes of very-large-scale integrated circuits (VLSICs). The manufactured products meet world standards (NORTON company) as evidenced by the growing volume of purchases of tooling in Europe and the United States. The plant that is friendly to our organization makes accessories that can deposit VLSICs on silicon wafers with a diameter from 75 to 300 mm (Figure 4).

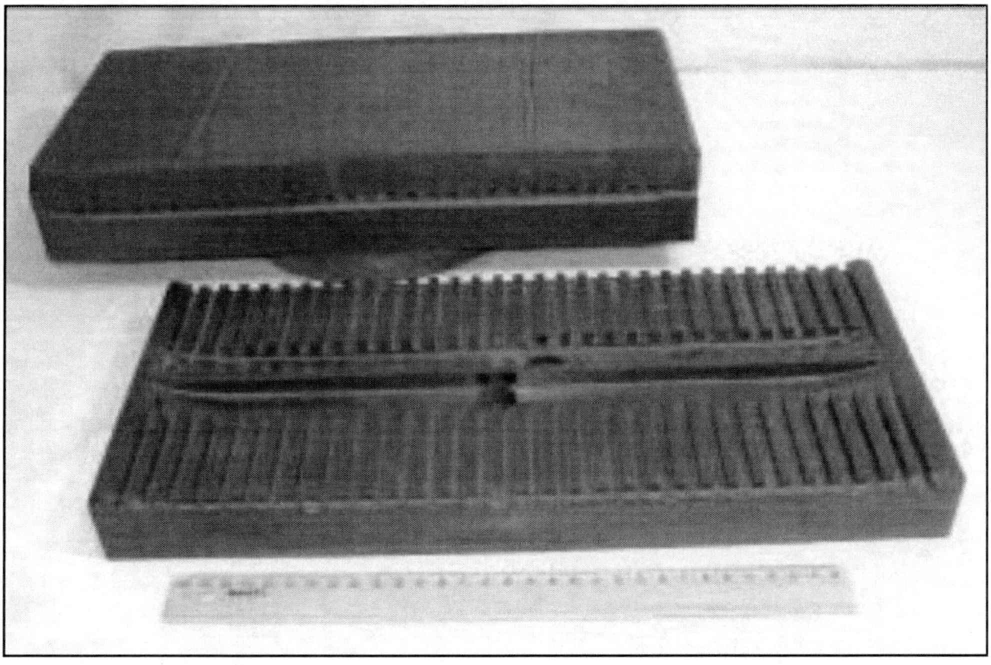

Figure 3. Elements of a cooled mirror.

The high thermal conductivity of silicon carbide (λ > 250 W/mK for sublimated SiC and λ ~ 490 W/mK for single-crystal SiC) enables the use of silicon carbide as a heat exchanger. Currently, the technology of manufacturing high-performance SiC heat exchangers for semiconductor laser diode arrays has been perfected. Thus, in close cooperation with industrial enterprises we have fabricated and tested samples of laser diode arrays (Figure 5) with an output power from one array at a level of 100 – 120 W.

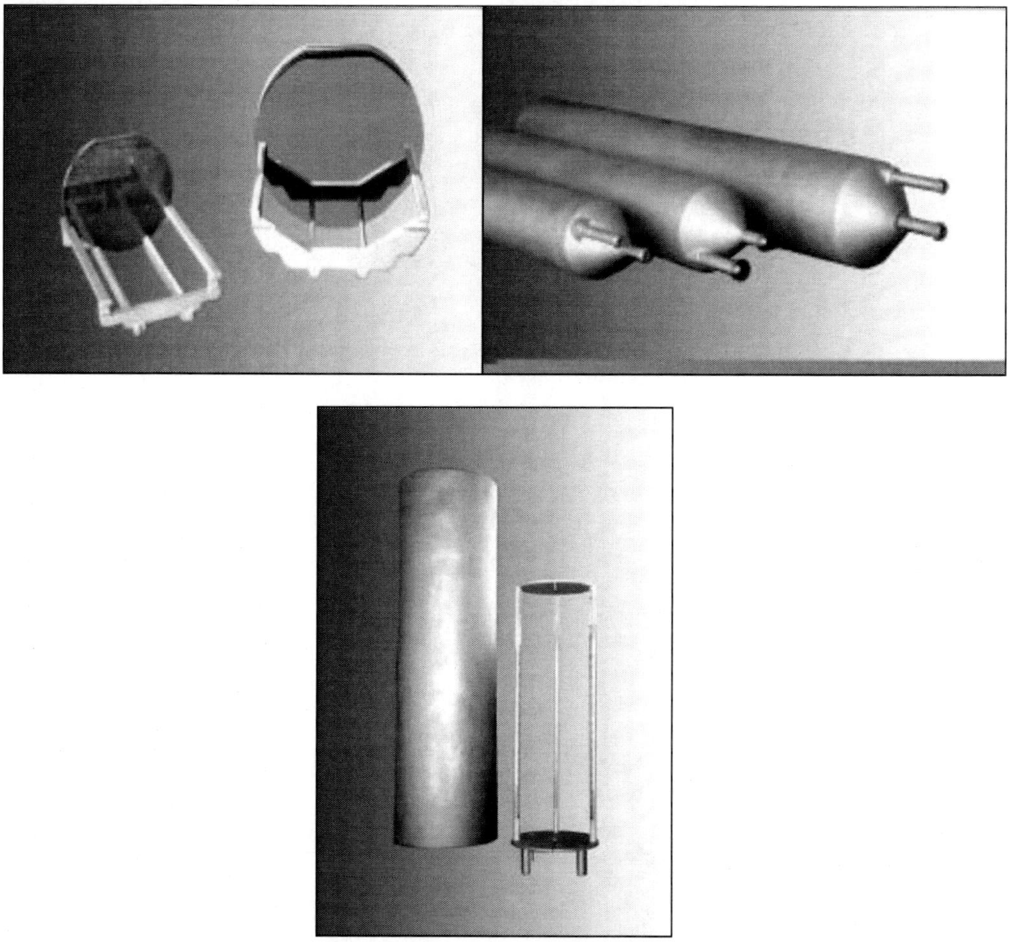

Figure 4. Accessories made of high purity silicon carbide for VLSIC production: (a) cassette for silicon wafers up to 200 mm in diameterб (b) reactor for annealing wafers up to 200 mm in diameterб and (c) reactor and cassette for wafers 300 mm in diameter (presented by Dr. G.I. Babayants).

The use of silicon carbide as a heat conductive material is also important because its thermal expansion coefficient is close to that of gallium arsenide (GaAs), a substrate of a semiconductor laser array, which prevents its cracking during brazing to the body of the heat exchanger. Studies are under way and positive results have been obtained with regard to radiation phasing on two-dimensional arrays of semiconductor laser diodes.

Currently, research is under way to develop technology for growing silicon carbide single crystals for microelectronic devices operating at high temperatures and fluxes of ionizing radiation.

Figure 5. SiC heat exchanger for an array of semiconductor laser diodes.

REFERENCES

[1] Apollonov V.V., Barchukov A.I., Karlov N.V., Prokhorov A.M., Shefter E.M. *Sov. J. Quantum Electron.*, 5, 216 (1975).

[2] Apollonov V.V. *High-Power Optics* (New York: Springer, 2014).

[3] Bochkov N.N., Gordeev V.F., Kolesov V.S., et al. Poverkhnost', *Fizika, Khimiya, Mekhanika,* (11) (1983).

[4] Didier Castel, Bertrand Calvel, Philippe Lamy, Kjetil Dohlen, Micharl Bougoin. Monolithic SiC telescope of the OSIRIS narrow-angle camera for cometary mission ROSETTA. Part of the SPIE Conference on Advanced Telescope Design, Fabrication, and Control, Denver, Colorado, July 1999, SPIE Vol. 3785 (1999).

[5] Michel Fruit – Axel Schindler, Thomas Hänsel. *Ion beam figuring of SiC mirrors provides ultimate performances for any type of telescopes.* IOM, Permoserstrasse 15.

Chapter 16

HIGH POWER/ENERGY LASER FOR WATER SURFACE CLEANING

Introduction

What is the best way to start our paper? "Deep water Horizon" (BP's operation) case in the Mexican gulf is the best example of manmade natural disasters. Today, the notion that "offshore drilling is safe" seems absurd. The Gulf spill harks back to drilling disasters from few decades past — including one off the coast of Santa Barbara, Calif. in 1969 that dumped three million gallons into coastal waters and led to the current moratorium. The "Deep water Horizon" disaster is a classic "low probability, high impact event" — the kind we've seen more than our share of recently, including space shuttle disasters, 9/11, Hurricane Katrina and an earth quake in Japan. And if there's a single lesson from those disparate catastrophes, it's that pre-disaster assumptions tend to be dramatically off-base, and the worst-case scenarios downplayed or ignored. The Gulf spill is no exception. Fire boats battle the fire on the oil rig "Deep water Horizon" after the April 21 terrible explosion.

The post-mortem is only the beginning, so the precise causes of the initial explosion on the drilling platform and the failure of a "blowout preventer" to deploy on the sea floor probably won't be established for weeks or months. But the outlines of serious systemic problems have already emerged, indicating just how illusory the notion of risk-free drilling really was, while pointing to some possible areas for reform. A "blowout" on an oil rig occurs when some combination of pressurized natural gas, oil, mud, and water escapes from a well, shoots up the drill pipe to the surface, expands and ignites. Wells are equipped with structures called blowout preventers that sit on the wellhead and are supposed to shut off that flow and tamp the well. "Deep water Horizon"s blowout preventer failed. Two switches — one manual and an automatic backup — failed to start it. When such catastrophic mechanical failures happen, they're almost always traced to flaws in the broader system: the workers on the platform, the corporate hierarchies they work for, and the government bureaucracies that oversee what they do. According to the study, 600 major equipment failures in offshore drilling structures of 80 percent were due to "human and organizational factors," and 50 percent of those were due to flaws in the engineering design of equipment or processes. With near-shore and shallow reserves of fossil fuels largely depleted, drilling has moved farther off shore, into deeper waters and deeper underground. The technology for locating oil and gas reserves and for drilling has improved, but the conditions are extreme and the challenges more formidable. This is a complex system. You've got equipment and steel strung out over a long piece of geography starting at the surface and terminating at 18,000 feet below the sea floor. So it has many potential weak points. Just as Katrina's storm surge found weaknesses in those piles of dirt — the levees — gas likes to find weakness in anything we connect to that source. It must be questioned whether energy companies and government agencies have fully adapted to the new realities. The danger has escalated exponentially. We've pushed it to the bloody edge in this very, very unforgiving environment, and we don't have a lot of experience. Disaster has several possible insights for the oil spill: one was that BP and other corporations sometimes marginalize their health, safety, and environmental departments. BP and other companies tend to measure safety and environmental compliance on a day-to-day, checklist basis, to the point of basing executive bonuses on those metrics. But even if worker accident rates fall to zero, that may reveal nothing about the risk of a major disaster. These things we are talking about are risks that won't show up this year, next year — it may be 10 years down the road before you see one of these big blowouts or refinery accidents.

That assumption — that catastrophic risks were so unlikely they were unworthy of serious attention — appears to have driven a lot of the government decision-making on drilling as well. One, published in 2007, estimated the "most likely size" of an offshore spill at 4,6 K barrels. Current, conservative estimate of the Gulf spill put its total at more than 80 K barrels, increasing at a rate of 5 K per day.

Why have we made such a detailed attention to the policy of BP`s "Deep water Horizon"? The contamination of large water areas (oceans, seas, lakes, and rivers) with petroleum products as a result of accidents and not so smart industrial activity of similar companies and officials around them is one of major problems of protection of environment. Any other contaminant can not be compared with petroleum on a basis of universal utilization, number of contamination sources and degree of effect on all components of environment. Through penetration of petroleum products into water, there are deep, frequently irreversible changes of its chemical, physical and microbiological and even global (redirection of the ocean streams) properties.

The following methods of disposal of petroleum contamination of water: natural, mechanical, physical-chemical, chemical, and biochemicals are widely used nowadays.

Natural methods demand to leave the oil alone so that it breaks down by natural means. If there is no possibility of the oil polluting coastal regions or marine industries, the best method is to leave it to disperse by natural means. A combination of wind, sun, current, and wave action will rapidly disperse and evaporate most oils. Light oils will disperse more quickly than heavy oils.

The mechanical methods are such: collecting of petroleum from a surface manually or with different installations. All these methods are effective for a limited time (from several hours to several days) only – the time period during which the thickness of a petroleum film is great enough. Contain the spill with booms and collect it from the water surface using skimmer equipment. *Spilt oil floats on water and initially forms a slick that is a few millimeters thick.* There are various types of booms that can be used either to surround and isolate a slick, or to block the passage of a slick to vulnerable areas such as the intake of a desalination plant or fish-farm pens or other sensitive locations. Boom types vary from inflatable neoprene tubes to solid, but buoyant material. *Most rise up about a meter above the water line.* Some are designed to sit flush on tidal flats while others are applicable to deeper water and have skirts which hang down about a meter below the waterline. Skimmers float across the top of the slick contained within the boom and suck or scoop the oil into storage tanks on nearby vessels or on the shore. However, booms and skimmers are less effective when deployed in high winds and high seas.

Use of dispersants break up the oil and speed its natural biodegradation. Dispersants act by reducing the surface tension that stops oil and water from mixing. Small droplets of oil are then formed, which helps promote rapid dilution of the oil by water movements. The formation of droplets also increases the oil surface area, thus increasing the exposure to natural evaporation and bacterial action. Dispersants are most effective when used within an hour or two of the initial spill. However, they are not appropriate for all oils and all locations. Successful dispersion of oil through the water column can affect marine organisms like deep-water corals and sea grass. It can also cause oil to be temporarily accumulated by sub-tidal seafood. Decisions on whether or not to use dispersants to combat an oil spill must be made in each individual case. The decision will take into account the time since the spill, the weather conditions, the particular environment involved, and the type of oil that has been spilt and many other parameters involved in that hard consideration.

Introduce biological agents to the spill to hasten biodegradation. Most of the components of oil washed up along a shoreline can be broken down by bacteria and other microorganisms into harmless substances such as fatty acids and carbon dioxide. This action is called biodegradation. The natural process can be speeded up by the addition of fertilizing nutrients like nitrogen and phosphorous, which stimulate growth of the microorganisms concerned. However the effectiveness of this technique depends on factors such as whether the ground treated has sand or pebbles and whether the fertilizer is water soluble or applied in pellet or liquid form.

Besides that, the biological activity of microorganisms strongly depends on the temperature of water. As for physical-chemical methods it is necessary to mention, first of all, application of various adsorbing materials (polyurethane foam, coal dust, sawdust etc.), however all these methods are labor-consuming and low efficient. Besides they require secondary reprocessing of adsorbents.

Chemical method is the method of petroleum removal with the help of chemical substances. The basic disadvantages of this method are high price and the fact that detergents are frequently more toxic for water microorganisms than petroleum.

The laser method of cleaning of water surface from a thin petroleum film is one of physical-chemical methods. Some time after penetration into water, the petroleum spreads on a surface of the water and forms very thin film (thickness several microns). This film cannot be forced to burn, since because of good thermal contact to a surface of water the film cannot be heated to temperature ensuring steady combustion.

The principle idea of a laser method consists in the following. The laser beam passes through a petroleum film and then is absorbed in thin layer of water. The water heats very fast up to the boiling temperature, and the forming vapor destroys a film, bursting it into small-sized fragments, which, mixing it up with the hot vapor, are decomposed quickly with formation of simple un-toxic substances.

The main advantages of a laser method consists of the following.

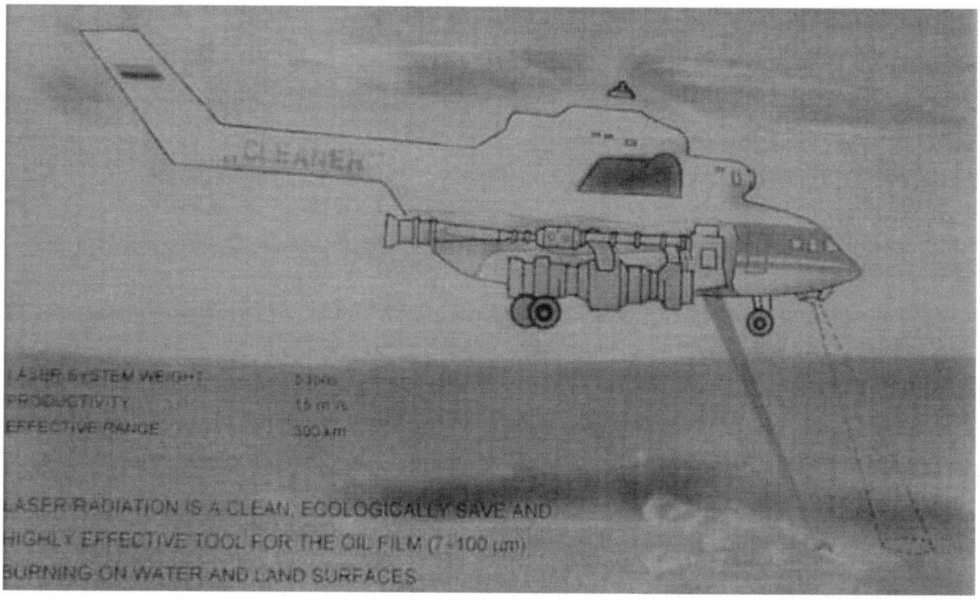

Figure 1. Mobile laser system for oil pollutions of water and land surface cleaning.

a) This method is a "fast-response" one since it does not require any special preparation; in emergency situations the time from the moment of obtaining the alarm signal to the beginning of the laser installation operation is determined only by the time necessary for arrival of the ship or the helicopter with the installation on board in a given area.
b) The method is contactless, i.e., does not require realization of preparatory or other activities in the oil spillage.

For realization of the proposed technology, it is possible to utilize different types of lasers, as continuous (power up to 250 kW), and high repetition rate pulse-periodical (average power of the same level with duration of pulses 100 ns and repetition rate up to 100kHz).

Theoretical estimations and the experiments have shown that as a result of laser beam action on a surface of a water and land, covered with film of hydrocarbon contamination, the following effects can be observed:

Evaporation and burning, and in a continuous mode the consumption of energy for 1 gram of vaporized liquid on the order of value surpasses the energy necessary for heating up to boiling temperature and evaporation of a film, which is explained by a heat consumption for heating water. In a pulse mode, the consumption of energy is per unit of vaporized mass of a liquid (film) approximately 5 times less, thus the process of film ignition starts more easily.

Knocking out the particles of polluting substance above a surface of water under action of pulse or powerful scanning laser radiation should be considered [1]. The physical mechanism of this effect is explained by sharp evaporation (boiling-up) of a thin layer of water under a layer of polluting substance. This process takes place whereas hydrocarbon polluting thin film has an absorption coefficient less than water. Knocked out particles of petroleum at a height up to 50 cm is possible. For this case, the energy consumption per unit mass of a raised liquid is sufficiently less than the energy necessary for evaporation and ignition (mode number one) or for sucking up and saving for future efficient usage (mode number two).

The character of a task to be solved superimposes certain conditions on operating characteristics of the laser device concerning both parameters of laser source, and concerning auxiliary systems.

Technological lasers of 10 - 15 kW power range that are widely used in production will have the output completely insufficient for liquidation of large scale contamination presenting the greatest danger according to the above-stated estimations. Such lasers can be utilized, at best, for improvement of a process engineering of cleaning under modeling conditions.

Besides, the character of this task dictates the impossibility or extreme undesirability of use of the stationary civil engineering service line (electric power network, water pipe, main gas line, etc.), in view of the requirement of a high self-sufficiency and mobility. The power unit should not limit mobility and thus provide total energy needed for all installations, the necessary reserve of expendables and fuel on board a complex. A capability of a fast redeployment from one type of a vehicle on another is also desirable.

In the present technical paper, the basic design concept of the mobile laser installation on basis of gas dynamic CO_2-laser is developed. The activities were executed in accordance with the working plan of OOO "Energomashtechnika"

In the first paragraph, the possibility of using of various types of lasers for solution of required tasks is considered and the selection of GDL is justified.

The second paragraph is devoted to selection and substantiation of basic performances of the laser installation and principal schematics of GDL.

In the third paragraph, the possible schemes of organization of air supply in the laser installation are considered. The selection of jet engines for the high quality power unit of GDL installation has been provided. The necessary volume of the selected jet engine adaptation/modification of experimental works is presented.

The fourth paragraph of the paper is devoted to the description of the design concept of GDL, and also laser installation as a whole. The pneumohydraulic schematics are described, the structure of the equipment, fuel tanks and operation control units are determined. The description of jet engine based CO2 GDL and detailed analysis of the laser installation components is also presented.

THE SUBSTANTIATION OF SELECTION OF LASER TYPE ACCORDING TO SPECIFIC PARAMETERS, OPERATION AUTONOMY, AND MOBILITY

There are a few of the most effective and scalable modern high power continuous/pulse-periodical operated gas lasers that should be considered for the claimed realization in the title of the paper tasks: electro discharge laser (EDL), gas dynamic laser (GDL), chemical laser HF/DF (CL) and chemical oxygen - iodine laser (COIL). We are not going to discuss details of high power high repetition rate P-P laser systems operation. Main results of oil film elimination with detailed consideration taken in the past can be summarized and reduced to paper format and presented here as follows:

Analysis of Applicability of Various Types of Lasers to the Task

The laser with output power up to 250 kW of continuous operation (CW) or high repetition rate pulse-periodical operation (P-P) during a few hours (minimum requirement) is required for realization of the above-mentioned task. Besides that, according to the physics of the process of destruction of a petroleum film, explained in the Introduction, laser radiation should be weakly absorbed by the petroleum film and should be absorbed effectively by water. In Table 1, the computational data for depth of radiation penetration for 4 types of above-mentioned lasers in petroleum and in water are presented.

Table 1. Laser radiation penetration depth for petroleum and water

The laser type	**Depth of penetration, micron**	
	Petroleum	**Water**
GDL, EDL	100 - 260	10
CL	26 - 200	0,8 - 40
COIL	50	2000

The range of values of depth of penetration for petroleum is the sequence of the fact that absorption coefficient of various grades of petroleum and water (for the CL case) in relation to wavelength in CL radiation spectral band is considered.

From Table 1, it is obvious that the COIL radiation is rather weakly absorbed by water and is rather strongly absorbed by petroleum. However, to heat a petroleum film by thickness ~100 microns through the absorption mechanism of COIL radiation is practically impossible because of strong heat transfer from a film into water through the heat conductivity. The conclusion is: the use of COIL to solve this task is impossible.

As the lengths of waves of radiation GDL and EDL are identical, the comparison of these two types of lasers is carried out on the basis of mass - dimension characteristics and other parameters that are listed in Table 2. The comparison is carried out for two real installations, which were developed and tested in the past to a sufficient degree, precisely - GDL of rated power up to 250 kW, and EDL with 20kW output power.

Table 2. The comparison of the parameters for GDL (250 kW) and EDL (20 kW)

#[1]	**The name**	**GDL**	**EDL**
1.	Power laser (max), kW	Up to 250	20
2.	Overall dimensions of the laser installation, m	4x2.4x2.4	2.5x3x3.5 + 2 x 0,8 x 0,8
3.	"Dry weight " (free of fuel), metric ton	5	18.6
4.	Specific, 1 kW of laser power dry weight, t/kW	Up to 0,02	0,93
5.	Specific, 1 kW of laser power, dry volume, cubic m/kW	0,1	1,5
6	Specific volume fuel consumption for 1 kW of laser power, cubical meter/hour/kW	0,02	0,01

The analysis of the data presented in Table 2 shows that GDL has decisive advantages in comparison with EDL. It is necessary to mention here that EDL specific fuel consumption is 2 times less than that of GDL. However, influence of this factor on the total complex weight will demonstrate a negative influence only for large operation time (5 hours or more). At the arrangement of a laser complex on board the helicopter, the operating time will not exceed 1 hour, therefore in this case smaller EDL specific fuel consumption will not take decisive value.

Thus, GDL and CL remain the only competing systems. The comparative analysis of these two types of lasers with reference to the considered task is stated below.

Specific Power of Laser Generation

GDL specific power, Wsp is about 20-35 J/g (C2H4+toluene); this value for CL is much higher and reaches 150 J/g. From this point of view, CL has a big advantage in comparison with GDL. But taking into account a few other parameters, one can say it is not the final conclusion for this particular story. Very important parameters for our task should be considered as well: wavelength, scalability of the system, technical maturity of the technology, safety, life time of hardware and so on. Up to now, GDL looks like the best system for the task under consideration - cleaning of water surface from petroleum films.

Wavelength of Radiation

GDL radiation wavelength, λ = 10,6 microns and CL wavelength, λ= 2,7 - 4,5 microns. From this point of view, CL has one more potential advantage over GDL, as the diffraction limited angle of divergence of the CL beam is 3 - 4 times less in comparison with GDL (at identical aperture). However, this problem requires further precise consideration.

a) Radiation extraction from the resonator.

In high-power lasers, extraction of laser beam from the resonator, where the pressure is much lower than atmospheric pressure, in external space through a rigid window is practically impossible for the reason of thermal destruction of such a window. For an extraction of such a beam out of installation, the aerodynamic window (AW) can usually be utilized. The supersonic gas stream cross-sectional to the beam direction fulfills the task of transparent boundary. As there are areas of various densities, gas sheaths limited by curvilinear surfaces in such flow, the beam transmission through this gas in uniformity introduces the distortions of a wave front resulting in increase of an angular divergence of a beam. These distortions do not depend on a wavelength as the dispersion of an index of refraction of gas is extremely small and therefore these distortions have the same absolute value for GDL and CL. However relative values of these distortions of a wave front set (i.e., reduced to wavelength) are for CL 3-4 times higher than for GDL, therefore from the point of view of an angular divergence of radiation, the advantage of CL in comparison with GDL is not so essential.

b) External optical system.

The optical systems GDL and CL for delivery beam energy to remote objects necessarily include external (in relation to the resonator) optical elements and systems intended for expansion of a beam with the purpose of reduction of its divergence. A principal component of a telescope is the main mirror, which diameter can reach the value up to several meters (depending on required power density on the target). The one piece of glass or metal mirror will be too heavy. For this reason, the main mirror usually consists a large number of small mirrors (facets), assembled in one unit. Every facet has its own adjusting device. The accuracy of relative positioning of these facets should not be worse $\lambda/4$, otherwise distortions of a wave front after reflection of a laser beam from this mirror will cause the essential increase of an angular divergence in comparison with its diffraction limit. This is $\lambda/4 \cong 2,5$ microns for GDL, $\lambda/4 \cong 0,75$ microns for CL. Therefore, in practice, required accuracy is much more difficult to ensure for CL.

This brief discussion shows that the efficiency of a CL external optical system can be much below that of GDL and, hence, reduces the advantage of shorter wavelength of CL radiation.

Besides that, for small delivery length of the beam energy down to the water surface, the angular divergence has no great importance, since the density of power on a water surface appears sufficiently high without application of an additional external optical system.

Length of Radiation Generation Zone

In the GDL working section (the resonator region), the transfer of oscillation energy from N_2 to CO_2 has taken place. The characteristic time of this process is relatively large. Besides, the relaxation time of oscillatory - excited molecules of CO_2 at their collisions with other particles is also large. Therefore, for full extraction of N_2 accumulated oscillatory energy, and transformation in its laser radiation, the length of a working zone in flow down direction should be 10 - 15 cm. Such large length provides certain freedom of selection of the resonator scheme (one pass or multipass, symmetrical or asymmetrical, etc.). Thus, the mirror operating surfaces are large enough, which result in a reduction of its thermal loading and distortion of a surface due to inhomogeneous heating.

In CL, the length of the beam generation section is small, less then 2 cm. This circumstance causes difficulties in optimization of the laser resonator and leads to increase of thermal loading of mirrors, which results in significant thermal distortion of mirrors and distortion of a wave front, i.e., to significantly increase an angular divergence of laser radiation.

Direct Exhaust of Used Gas to Atmosphere

If GDL or CL are placed on the ground or on board a flying vehicle (plane or helicopter), there is a problem of exhaust of spent gas to the atmosphere. In the case of GDL, this problem can be solved rather simply by use of the diffuser, since the operating pressure in the zone of the resonator is rather high. For reaching a maximum degree of pressure recovery in the diffuser, it should have a very special geometry.

In the CL case, the operating pressure in the resonator is much lower in comparison with GDL, therefore the direct exhaust to the atmosphere with the help of the diffuser is impossible. It is necessary to additionally utilize the gas or water vapor ejector to ensure necessary pressure in the output section of the diffuser. Such ejector requires a large additional gas flow rate, which results in fundamental complexity of the laser installation design and significantly decreases specific power of generation, i.e., in this case, one of the main CL advantages is lost in comparison with GDL.

Toxic Characteristics

The degree of CL danger from the point of view of toxic characteristics of working components and exhaust gases is much higher in comparison with GDL. In CL, for creation of a working mixture, the extremely toxic substances containing fluorine should be utilized. Besides the exhaust, CL contains a significant quantity of fluorine hydride HF, which is a super toxicant. In this connection, with use of CL in a system of laser cleaning of water areas, it will be necessary to ensure special safety measures, which will cause significant complication of the installation design and rise of its price.

In the case of GDL, exhaust gas is ecologically clean (N_2 + CO_2 + H_2O). However, CO, carbon monoxide, which is a toxic gas, can be used as a fuel for GDL. Hence, a system of storage and supply of CO should ensure absence of its leakage. On the other hand, CO is lighter-than-air; therefore it floats in the atmosphere and is fast blown away by wind. If used as fuel, GDL will be used as liquid hydrocarbon fuel (benzole, toluene), the system of its storage and supply does not require special measures to secure the ecological safety.

Absorption of Laser Radiation by Atmosphere

The wavelengths of GDL and CL radiation are within windows of atmosphere transparency. These windows are usually determined for vertical transmission of beams through the atmosphere, and for this case power dissipation on track is ~20% for GDL and 25% for CL. For vertical tracks of length 50 - 150 m, which are of interest with reference to a considered problem, the radiation power dissipation can be neglected.

Substantiation of Selection Gas Dynamic CO_2-Laser

Proceeding from the comparison of various type lasers explained in the previous paragraphs, it is possible to make a conclusion that the most suitable laser for development of the installation for disposal of petroleum films is GDL:

Simplicity of Design

From the design point of view, GDL components are the most simple in comparison with the considered lasers. The combustion chamber operates on usual components, for example, kerosene + air. The design of such chambers is very mature; there is a wide experience of their operation in various technical devices. The nozzle unit of GDL is made from heat resisting steel and can work a long time without forced cooling. Wide zone of the media population inversion behind the nozzle unit (10-15 cm) stipulates simplicity of selection of the scheme of the resonator for obtaining laser generation with maximum efficiency. Thus the characteristic size of a mirror of the resonator is approximately equal to length of a zone of generation, i.e., 10-15 cm, therefore manufacturing of these mirrors does not entail serious technological difficulties.

The capability of GDL activity at a high pressure of gas in the resonator in front of the nozzle unit (up to 3 MPa and more) provides a straight line of exhaust of spent gas in the atmosphere with the help of the supersonic diffuser of a special design. Considering such capability, all remaining lasers are deprived. The aft ejector is necessary for maintenance of a direct exhaust of these lasers, which requires additional gas flow rate, which complicates the installation and reduces specific energy of radiation.

It is necessary to also note that GDL output power is rather weak and varies at change over a wide range of parameters of gas (pressure, temperature, chemistry) in the combustion chamber, i.e., GDL is not critical to accuracy of a task and maintenance of an operational mode. One more virtue of GDL is the small time (some seconds), necessary for start.

Thus, GDL is the rather simple, reliable and flexible tool ensuring high output power of laser radiation.

Mobility

From all considered types of lasers, GDL apparently has the least "weight" and "volume" of 1 kW of laser power. It is rather simple to place GDL on any vehicle, whether it will be an airplane or helicopter, ship or railway platform, etc. Such GDL mobility allows to help solve many tasks, including those connected with disposal of petroleum films.

Selection of Propellant Components

One of the specific singularities of working process of the CO_2-laser is that the working mixture of gases should be contained in a receiver rather than a small quantity of water vapor (no more than 5-6 volume %). On the other hand, the operation temperature of gas in a receiver should be high enough (1300-1600 K). It superimposes limitations on selection of propellant components (fuel and oxidizer). From the point of view of maximum specific energy of radiation (J/kg), optimum components are gaseous at standard conditions with damp (CO) and liquid nitrous oxide (N_2O), thus additional nitrogen (or air), necessary for working process, moves in a receiver from a separate source, and the water vapor in quantity 1,5-2% is formed by burning a small quantity of hydrogen or alcohol in air.

Wavelength of Radiation

The analysis of the data presented in tab. 1 show that for the solution of a problem of disposal of petroleum films, the optimal wavelength of radiation is that of the CO_2-laser, as this radiation is weakly absorbed by the film and is strongly absorbed by water. From the point of view of a diffraction limited angular divergence of radiation, the CO_2-lasers (EDL and GDL) lose for CL. However with reference to a considered problem, the value of a divergence is not critically important. The estimations show that the length of a beam from the laser source down to a surface of water ~ 50-100 m and power of radiation up to 250 kW, precision focusing of laser beam is not required for the reason of obtaining required density of power in a spot on the water surface. It is possible to utilize an unfocused or partially focused beam. Therefore, with reference to a considered problem, the relatively large wavelength of GDL radiation is not the factor of insufficiency.

The reasons explained above allow making an unequivocal conclusion for the selection GDL as the laser radiation generator for the mobile installation intended for disposal of a petroleum film on a surface of water. At the estimated power of radiation, GDL is preferable in weight and volume factors in comparison with EDL, HCL and COIL: it does not require the electric power used to utilize low toxic fuels, and provides the direct exhaust of utilized gas to the atmosphere. GDL is very simple in control. It is not critical to changes of working parameters in a sufficiently broad range, and it is convenient in operation. The important circumstance is that we had in our hands a very reliable and operable GDL with output power 100 kW, which in our days is a very effective tool for use in an intensive research program. In particular, the design of the aerodynamic window permitting extraction of a laser beam from the resonator zone to the atmosphere, without application of transparency for the working wavelength materials, has been effectively developed.

Selection and Substantiation of Basic Performances of the Laser Installation and GDL Principal Diagram

In the previous chapter, the comparison of various types of high power lasers was carried out from the point of view of their application for the solution of this task, where it was shown that the most appropriate type of laser is CO_2 - GDL. However, there are different types of GDL therefore it is necessary to choose the optimal version, and also to develop the general concept of the power installation for optimal supply of working components in GDL.

In the present chapter, the analysis of the various GDL schemes, their advantages and disadvantages are given from the point of view of application in the considered installation, the selection of propellant components for GDL is justified from the point of view of their power efficiency, production and toxic characteristics, and ecological safety.

The last part of the chapter is devoted to the substantiation of the GDL characteristics working on the chosen components. For this purpose the calculation results, which have been carried out according to [2], are permitted to define the specific power of laser radiation at the given initial parameters, such as fuel chemistry, temperature and pressure in front of the nozzle block, expansion ratio of the nozzle etc. are used. The calculation results and computer analysis are shown.

To define the main dimensional characteristics of the laser and its flow organizing parts, it is necessary to define the basic level of laser radiation power, which, in turn determines the total fuel consumption and remaining geometrical characteristics of the laser.

THEORETICAL SUBSTANTIATION OF POWER LEVEL

The basic capability of disposal of petroleum contaminations from water surface by means of irradiation of a petroleum film by a high flux of laser radiation has been proved experimentally and presented in our publications. However, the published data give only the general representation about the basic possibility of using such method. It is obvious that combustion is a complex multi-parameter process and, if we talk about operational use of the mentioned principles in the particular technical device intended for practical use, the optimization of parameters of the installation is necessary. The important value in this case is the efficiency, which is possible to understand as the area of cleaned water surface divided by mean energy, used on clearing, (m_2/kW). For optimization of parameters of the installation it is necessary to conduct a cycle of research working on the study of evaporation, ignition and combustion processes of petroleum used for the above-mentioned method. The experimental research of the given problem, even without qualitative understanding of involving processes, represents a very difficult task in connection with the vast quantity of optimization parameters, and requires heavy material costs and time. In this connection, the preliminary development of the theoretical model representing the disposal process of petroleum film is desirable.

The second phase [3] of the process is presented, the various interaction mechanisms of a laser beam with a petroleum film on a water surface are theoretically considered. The following mechanisms are included:

- laser heating active absorbing, translucent and transparent films;
- an explosive boiling-up and conditions of process efficiency;
- evaporation and vaporization.

The basic conclusions of the technological approach are the following.

1. For reaching temperatures of evaporation of the film, the necessary time of radiation action is about 100 nanoseconds at laser power 100 kW and the diameter of an irradiated zone is about 30 cm. Temperature of the water layer under the surface of film can essentially exceed the boiling point. Such situation arises due to superior velocities of laser heating (up to several millions degrees of Celsius per one second) and inertial character of heat transfer process. At such temperatures the water passes in a metastable state with active outgassing. Under the surface of a film, there will be microexplosions of air - steam bubbles. As a result, there will be a separation of a film from the surface of water, after breaking the outflow of drops of petroleum taking place in the open air.

2. The scanning velocity at power of the laser of 100 kW and diameter of a 30 cm spot ought to be in the range of 10-12 m/s. Thus, the velocity of cleaning of a surface 18000-24000 m^2/hour can be provided.

It is necessary to note that all calculations are fulfilled for thickness of a petroleum film of about 100 microns, which is a rather large value, and is observed only in the initial moments of formation of oil spillages. For much thinner films, the velocity of cleaning will be increased proportionally.

Experimental Research

For experimental confirmation of basic physical principles included in the basis of the Project, the participants carried out preliminary experimental research of interaction of laser radiation with films of various petroleum types on a surface of water. The research works were conducted within the framework of financing activities.

During our experimental research works, the GDL developed by our team and prepared for this task had been used. The optical scheme of realization of experiments is presented in Figure 2.

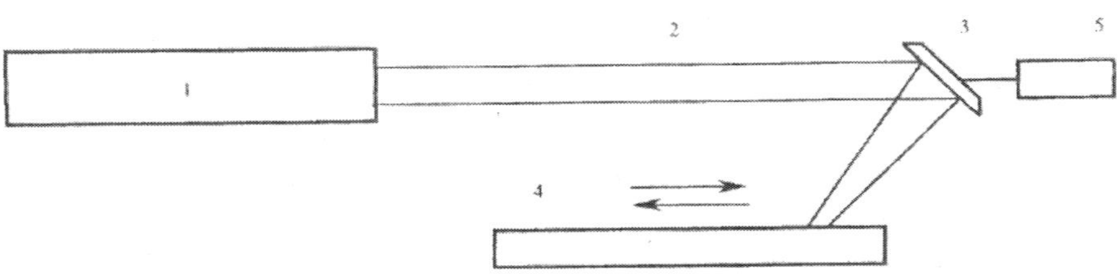

Figure 2. Optical scheme of oil film interaction with laser beam.

The GDL 1 generates radiation 2 with a wavelength 10,6 microns. The radiation is directed on a concave focusing mirror 3 and, being reflected, impacts on a surface of test heterogeneous structure in a cuveta 4. The mirror realizes periodic oscillations with the help of the mechanical device 5, thus the laser beam scans on a surface of a dish. Diameter of laser beam cross section on a surface of a dish changed from 15 cm to 30 cm. The laser radiation power was within the limits of 100 kW.

In Figures 3, 4 and 5 the results of experimental research works are presented, on which it is possible to make following preliminary conclusions.

1. The experiments testified to the possibility of using of the laser installations for cleaning of aqueous medium from contaminations of petroleum and its processing products.
2. For definition of optimal parameters of the laser installation and operation modes in practical conditions, it is necessary to select experimental conditions of the working regime according to the special program approved by a potential customer.
3. For necessary reserve, the radiation power of GDL of 100 kW for the installation to be developed according to the present consideration is chosen.

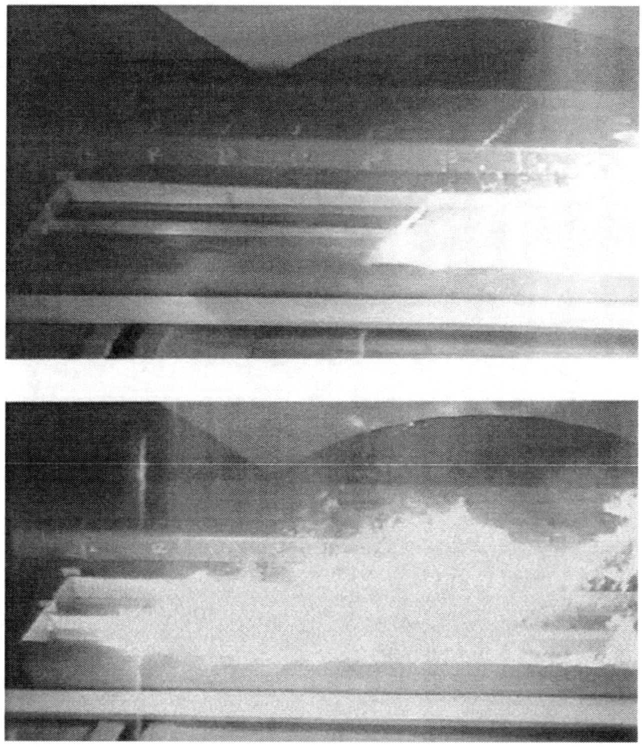

Figure 3. Intensive combustion of petroleum contamination under laser radiation.

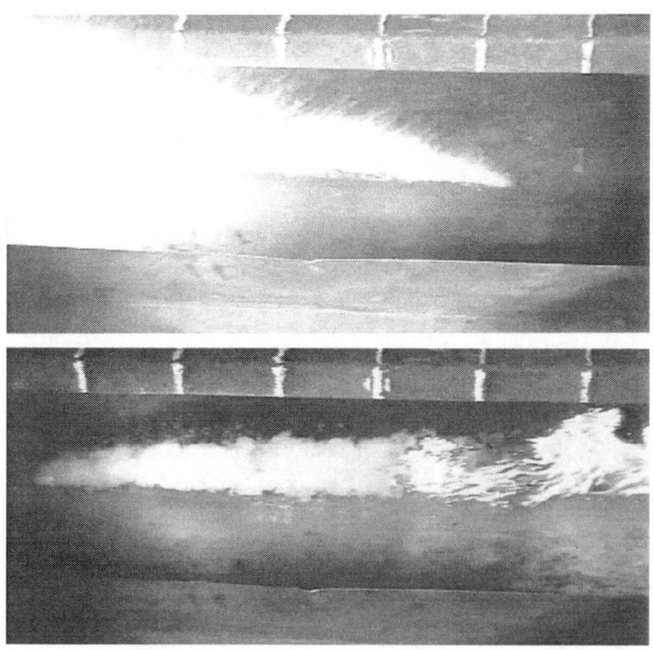

Figure 4. Combustion of kerosene under laser radiation.

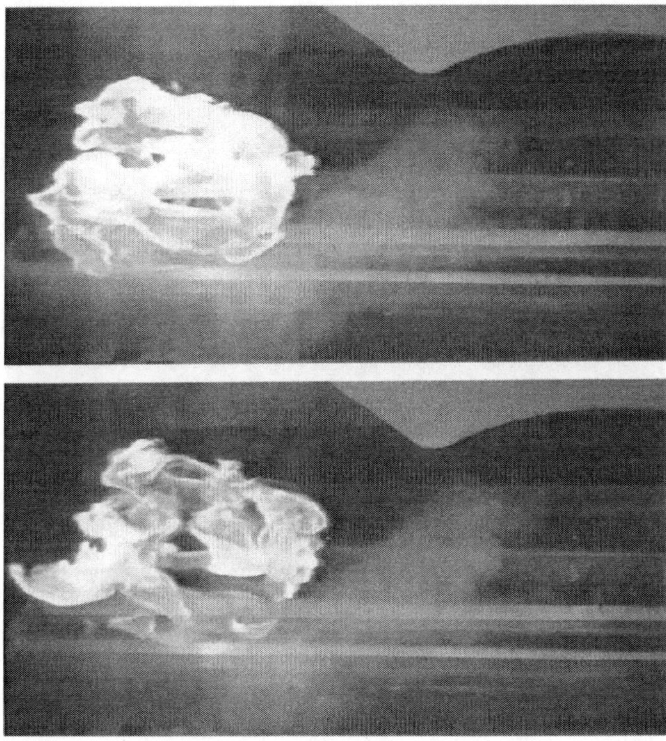

Figure 5. Combustion of engine oil under laser radiation.

Conventional Scheme of GDL

The principle of GDL operation [4] is based on non-equilibrium fast expansion of the heated working gas in the supersonic nozzle, at which there is a partial freezing population of oscillatory levels of molecules CO_2 and N_2 and formation of population inversion of upper and low laser levels. The heating of gas to an operation temperature 1300 –1700K at pressure 2-3 MPa is conducted in the combustion chamber through combustion of chosen fuel – oxidizer components. The transformation of the accumulated oscillatory energy of gas to laser radiation has taken place in the optical resonator located down the nozzle. After the resonator, the working gas passes through the supersonic diffuser and will be exhausted in the atmosphere.

The fastest gas expansion takes place near the critical zone of the nozzle at the length equal to the value of one order of magnitude of the critical cross-section size. Therefore, for fast freezing of oscillatory energy, it is necessary to use nozzles with a small size of critical cross-section (<1 mm). On the other hand, the pressure of gas after the nozzle (i.e., in the resonator region) should be small enough (<100 mm Hg) to reduce to a minimum the losses of oscillatory energy due to collision relaxation. It has been solved by use of nozzles with large expansion ratio (area ratio of output to critical cross-sections $A/A^* \geq 20$, that corresponds to the value of a Mach number M 4).

For effective transformation of the accumulated oscillatory energy to laser radiation, the amplification of a small signal during the pass the length of an active zone should be large enough to compensate optical losses (removal of radiation, absorption in mirrors, dissipation in medium) and to ensure a high enough radiation power in the resonator. In GDL of very low powers (tens W) is possible to use the so-called mononozzle, which represents the flat (rectangular) nozzle with critical cross-section as a narrow slot. However, high-power GDL implementation of the mononozzle is practically impossible because of its large length (1 m and more); arising thermal deformations result in essential to nonuniform width of critical cross-section, h *, down to full blocking of critical cross-section in some zones. Besides, the mononozzle does not receive enough large transversal size of an active zone, which causes certain problems at resonator selection. Therefore, in high power GDL, the so-called nozzle unit representing a set of a many relatively short (~100 mm) slot-type nozzles is used. These nozzles were formed by contoured nozzle blades, made from thermal resisting metal. Such design of the nozzle unit allows changing a wide range width and height of an active zone. Besides the nozzle unit provides a uniform density field of gas in an active zone, which allows for realizing a small angular divergence of GDL radiation.

In view of such advantages of nozzle blocks, this technology is widely used for GDL, and therefore it is now possible to name such GDL systems as the " GDL of traditional scheme". The radiation power of such a GDL is currently up to 250 kW and more for continuous and P-P radiation modes of operation [5, 6].

GDL with a Set of Axisymmetrical Nozzles

The nozzle unit of the conventional scheme has the disadvantage of being complex in manufacturing, and is subject to influence of thermal deformations. There is an alternative approach to the development of nozzle units, which consists of the following. Instead of two-dimensional nozzles, axisymmetrical nozzles with small diameter of critical cross-section are used. The supersonic part of such nozzle can be made conical or contoured. In this case, the nozzle unit looks like a plate, where the numerous small-sized nozzles are located as close as possible to each other. The manufacturing of such nozzles is possible since they are not subject to thermal deformations. With an appropriate arrangement of these nozzles on a holding plate (quantity of rows and quantity of nozzles in a row), it is possible to have an output cross-section of the nozzle unit with the required configuration.

However, such nozzle unit has at least two disadvantages caused by the presence of "empty" zones between the nozzles, since the complete filling of output cross-section by round nozzles is impossible.

The first disadvantage consists of the empty zones, which are the sources of shock waves in a supersonic flow. That wave results in a dense inhomogeneous structure of gas in a zone of the resonator and, hence, leads to increase of an angular divergence of laser radiation.

The second disadvantage consists of the effect, which creates the viscous vortices flows in empty zones, which results in reduction of total pressure in the main flow. In turn, the reduction of total pressure has negative consequences on the supersonic diffuser operation, which can be appreciated as reduction of pressure recovery coefficient. In this case, in order to direct the exhaust of working gas to the atmosphere it is necessary to raise combustion-chamber pressure, which results in the decrease of output power of laser radiation.

For these two reasons, GDL with the nozzle unit as a set of axisymmetrical nozzles is not widely used at the noticeable production scale.

GDL with Mixing of Working Gases in a Supersonic Flow

In the case of traditional GDL the simultaneous heating of all components of a working mixture - N_2, CO_2, H_2O has taken place in the combustion chamber. However, initial reserve of laser energy contains, primarily in oscillatory - excited molecules N_2, which is 90% of the gas mixture. The role of CO_2 and H_2O consists only of transforming accumulated N_2 oscillatory energy to laser radiation.

The pure nitrogen has a long oscillatory relaxation time stipulated by collisions of molecules N_2 among themselves. As the result, the oscillatory levels N_2 are easily frozen at fast expansion of the working mixture in the nozzle. However, other components of a mixture (CO_2 and H_2O) considerably accelerate a collision relaxation N_2 and this reduces a reserve of oscillatory energy.

To increase the reserve of oscillatory energy, it is desirable to raise temperature in the combustion chamber. However, if the traditional GDL temperature is limited by value about 2300K; the dissociation CO_2 begins at higher temperatures, and the efficiency of conversion of the accumulated oscillatory energy in laser radiation decreases. In contrast to CO_2 the molecular dissociation N_2 begins at temperature ~ 4000 K.

The presented ideas have resulted in the new GDL scheme, which consists of the following: the pure nitrogen, heated up in a prechamber, expands in the supersonic nozzle, and at the exit of the nozzle, mixed with cold CO_2 and H_2O. Such scheme has three basic advantages for the usual scheme GDL: a) temperature of nitrogen in a prechamber can be essentially increased, b) more effective freezing of oscillatory energy of nitrogen is reached, c) energy loss for CO_2 and H_2O heating is not necessary.

In view of the indicated advantages, mixing GDL can ensure the increase of specific power of radiation many times in comparison with the GDL conventional scheme. A main problem in mixing GDL is to ensure fast intermixing hot N_2 with cold CO_2, but the disturbances of a supersonic flow N_2 should be kept at minimum.

Some solutions to this problem are possible. Technically, it is much simpler to organize mixing of CO_2 and N_2 in wakes. In this case, the flow of CO_2 moves in parallel to the flow of N_2, in this case two sets of supersonic nozzles are used - one set for N2, the second for CO_2. An advantage of such scheme of supply is that arising at interaction of wakes, disturbances are minimal in comparison with other schemes of intermixing.

The basic disadvantage of this scheme is that the mixing of wakes is rather slow, therefore for effective work of the laser the working section (from the nozzle exit down to the diffuser inlet) should have sufficient length, which results in large relaxation losses and reduction of a pressure recovery coefficient of the diffuser. For these reasons, the scheme of GDL with intermixing of supersonic wakes is apparently not effective.

The theoretical analysis and the experimental data show that the GDL schemes are more effective, in which the injection of cold CO_2 is made near the critical cross-section of the nozzle in its transonic part. In experiments with heating N_2 in an electric arc plasma-generating device, a record supply of CO_2 was received for GDL value equaling a gain of 3% cm^{-1}, which exceeds the maximum value of a gain for usual GDL almost to an order of magnitude. However, specific power of radiation was insignificant, and is connected to a small scale of the installation.

Despite the basic promising character of mixing GDL, there are a lot of problems connected with their practical realization.

One of the most important problems is obtaining hot nitrogen at temperature 3000-4000K. Application of an electric arc plasma-generating device for high-power mixing GDL is unreal, since the consumption of nitrogen is 10 kg\sec and more required electrical power exceeds 100 MW. It is possible to use this for obtaining hot nitrogen special fuels, however all of them are strong toxicants.

As a source of hot nitrogen, other devices can be considered, in which combustion of metal in air takes place. Because of this strong exothermal reaction, hot nitrogen and oxide of metal as particles are formed, which are extracted from the formed flow with the help of a cyclone separator. However, the necessity of a very high degree of a nitrogen stream cleaning from particles makes this problematic.

Another important technical problem is the development of the nozzle (selection of a material, development of a robust design, which would maintain without destruction large heat flows in the region of critical cross-section.

The supersonic diffuser causes one more important problem. Laser translation temperature of working gas in an active zone should be sufficiently low, whereas temperature in a prechamber should be relatively high. For the coordination of these two conditions, it is necessary to use a supersonic nozzle with large Mach numbers (M=5 and more). However, the higher Mach number is, the lower is pressure recovery coefficient for the diffuser. Therefore, for a direct exhaust of gas in atmosphere it is necessary to develop special diffusers, thus it can appear that such task is impractical without use of a fodder ejector. However, consumption of inducing gas sometimes exceeds the consumption of induced gas, and in this case, specific laser power designed on a summarized gas flow rate considerably decreases.

From what is explained above, it is possible to make an unequivocal conclusion: taking into account problems connected to development of the high power mixing laser, and also the absence of the prototypes of such lasers for the solution of a task of disposal of petroleum film on water surface, it is necessary to choose GDL of conventional scheme.

Selection of Fuel for Mobile GDL

Currently, the analysis of using various fuels to support operations of the mobile CO_2-GDL intended for disposal of petroleum film on a surface of water is being carried out.

CO_2-GDL provides the laser radiation generation in a continuous/P-P mode of operation with radiation wavelength of 10.6 microns. A basic physical principle of operation gas dynamic of the CO_2-laser is the fast expansion in the supersonic nozzle of a mixture of gases (CO_2, N_2, H_2O) preheated up to temperature 1500K- 2000K. In a supersonic flow with Mach number M = 4 - 5 temperature of gases down to values 300K – 350K, necessary for population inversion of laser levels obtaining of a molecule CO_2. For nozzle operation at direct exhaust in the atmosphere after pressure recovery in the diffuser (the static pressure in a zone of generation is equal 5 10^3-1 10^4 Pa), the nozzle inlet pressure (stagnation pressure of a flow) equaling 2-3 MPa is necessary. At small transversal and longitudinal sizes of the supersonic nozzle (the height of critical cross-section is equal 0.3 - 0.5 mm, and length of the nozzle 20 - 40 mm) characteristic time of the gas flowing through the nozzle appears small and comparable with time of oscillating relaxation of molecules of a mixture of gases, i.e., oscillating - oscillating exchange (*V-V* process) and oscillating - translation relaxation (*V-T* process) has taken place in nonequilibrium mode. Because of non-equilibrium processes in oscillating exited molecules at certain gas mixture ratio and flow parameters in the supersonic nozzle, the inversion population of the upper laser level of molecules CO_2 (0001) is formed. The molecules CO_2, are light generating molecules (laser radiation is realized on transition 00^01 -10^00), N_2 molecules are donor molecules transmitting oscillating energy to the upper laser level of molecules CO_2 (00^01), the molecules H_2O are intended for population of the lower laser level of molecules CO_2, (10^00). The optimal ratio of mixture of gases for an effective work of gas dynamic CO_2-laser should contain CO_2 = 0.1, N_2 = 0.89, H_2O = 0.01 (volumetric fraction). It is necessary to note that the increase of a volumetric vapor fraction of (H_2O) results in decrease of the laser radiation power with other things being equal.

For mobile GDL, laser radiation, ensuring high power in CW/P-P modes, a basic method of thermal excitation is the combustion of such fuel components, which as final products give necessary components of a laser mixture at relatively high temperature. The optimal mixture ratio of gases at high temperature of mixture can be obtained, for example, at burning of carbon oxide (CO) and hydrogen (H_2) in air or other oxidizer (N_2O, N_2O_4) with subsequent use of ballast nitrogen.

The specific requirements to the analyzed mobile installation impose some limitations on use of this or that fuel, oxidizer and ballast gas. The main difference of this installation from existing lasers is the requirement of large duration of operation.

For processing the greatest possible area of the polluted basin, the laser installation should have an operating time appropriate for operation capabilities of a vehicle, on which it is secured. For the helicopter, this time is $t \sim 30$ min. The carried estimations of interaction of a beam with a water surface covered with thin petroleum film have shown that the power losses by heating water can be reduced through fast scanning of a surface by a beam of power $W \sim 100$ kW. These requirements allow estimating the total fuel and ballasting gas necessary for mobile GDL operation. The specific power (W_{out}) existing homogeneous CO_2-GDL, as a rule, does not exceed 10 kJ/kg. Thus weight of fuel and ballast gas for such GDL operation should be not less than $m = Wt/W_{out} = 18000$ kg. This value essentially will increase at the account of weight of the equipment for storage of working components. Apparently, such large weight is unacceptable for laser arrangement on a flight vehicle.

At the same time, the main part (80 95%) of weight of design components used for creation of a propulsive mass GDL, is the weight of an oxidizer and ballast gas. Thus, use of free air as an oxidizer and a ballast gas is a unique opportunity to guaranty this installation operation. As CO_2-GDL fuel, carbon monoxide is usually used with the hydrogen addition. Carbon monoxide allows receiving high values of specific power of radiation in comparison with liquid fuels, such as toluene, benzole, kerosene, but the weight this fuel together with a system of storage is rather high. Weight CO is 13 - 15% from the total mass of spent components, i.e., in our case 2000 - 3000 kg. Weight of a storage system of such quantity of gaseous carbon monoxide is ≈7000 kg, and summarized weight fuel and system of storage m_Σ = 9000 - 10000 kg. In case of use cryogenic state carbon monoxide summarized weight m_{Σ} can be reduced approximately twofold, but it considerably surpasses appropriate weight for liquid hydrocarbon. Weight of liquid hydrocarbon is 4 5% from total mass of spent components, and the tanks for storage are much lighter than for carbon monoxide, as the hydrocarbons are stored at atmospheric pressure. In our case, weight of liquid hydrocarbon is 700 - 900 kg, and summarized weight fuel and system of storage m_{Σ} = 1000 - 1500 kg.

- Taking into account the above-mentioned reasons, possible fuels in the present operation are benzole, toluene and kerosene, and as an oxidizer and a ballast gas (instead of nitrogen) free air compressed by the compressor.
- As it was mentioned above, for normal operation of supersonic nozzles of the laser, it is necessary to ensure high total pressure of a stream (P_0 = 2 - 3 MPa). The aviation compressors can ensure a compression ratio up to P_0/Pa = 20. Here P_0 is stagnation pressure of a working mixture in a GDL channel, Pa atmospheric pressure. In all further calculations the value of stagnation pressure of a flow P_0 = 2 MPa and maximum values, the appropriate Mach number of the nozzle M = 4.7 with the area ratio of an exit of the nozzle to the critical cross-section $F/F^* \approx 25$ was accepted.

Comparing various fuels, the basic parameter is the power efficiency, namely the specific power of laser radiation W_{out} (power of laser radiation referred to summarized mass flow of laser components), achievable at use of the given fuel. In this connection, the analytical-theoretical research of influence of various parameters on value W_{out} was carried out.

Temperature and chemistry of a working mixture formed at combustion of various fuels in air were determined by thermodynamic calculation. The necessary for calculation enthalpy of initial components was taken from [4]. For the reason of uncertainty of efficiency of the compressor (approximately 0.75-0.83) in a main part of executed calculations, temperature of inlet air in the combustion chamber was assumed to equal appropriate temperature at adiabatic compression up to value

P_0/Pa = 20. This temperature calculated according to the following formula

$$T = T_a \cdot \left(\frac{P_0}{P_a}\right)^{(\gamma-1)/\gamma},$$

where T_a - temperature of free air, γ - isentropic exponent. At adopted values T_a = 298K, γ= 1.4 temperatures of inlet air in the combustion chamber is T ~700K. The enthalpy, appropriate for this temperature of air is equal to I = 400 kJ/kg.

At thermodynamic calculation the thermal losses, which depend on a design of the combustion chamber and beforehand are not known, were not taken into account. In this connection, in the subsequent calculations of flow of a working mixture in the nozzle block inlet temperature *(K)* was taken equal to 0.95 of temperature obtained in thermodynamic calculation.

The thermodynamic calculations have shown that the working mixture contains the investigated temperature range (or oxidizer-to-fuel ratios) only necessary for operation CO_2 - GDL components (CO_2, N_2, H_2O) and oxygen (O_2), as the content of remaining components is insignificant. Therefore, in further research it was supposed that the working mixture consists only of carbon dioxide, nitrogen, water and oxygen. In the paper, the calculations of non-equilibrium flow of a mixture CO_2 - N_2 - H_2O - O_2 in the supersonic nozzle are carried out as well. The kinetic model describing the population change of separate oscillating levels of molecules CO_2 and N_2 at processes of *VT*- and *VV*- of exchange was used. All oscillating levels of a molecule CO_2 with characteristic temperatures $\theta < 5000$ K were taken into account. It was supposed also, that the velocity of exchange of oscillating energy to a Fermi-resonance considerably surpasses a velocity of all remaining *VT*- and *VV*- processes. Thereof, the modeling system of levels was introduced, some of which have a large degree of degeneration and correspond to several levels in a full system of levels of a CO_2 molecule. As a result of calculations, the population inversion and gain coefficient of the nozzle unit outlet was determined. The calculations of specific output power of laser radiation were conducted at the following input data.

1. The working mixture of gas dynamic laser is formed at combustion hydrocarbon fuel in air. The stagnation pressure of a flow equals 2 MPa, total temperature equals 1600K.
2. The nozzle unit of the installation consists of flat contoured nozzles with the height of critical cross-section 0.8 mm and expansion ratio 25.
3. The sizes of exit cross-section of the nozzle unit: length - 1600 mm, height - 130 mm.
4. The size of a cavity of the resonator along the flow direction - 200 mm.
5. The three pass unstable resonator with a multiplication factor $M = 1.4$. A reflection coefficient of mirrors $r = 0.98$, the losses on dissipation make $\beta = 0.03$ m^{-1}.

The contour of a supersonic part of the nozzle was obtained as follows. The flow of gas with an isentropic exponent equaling 1.35 in the contoured nozzle with an angular point was calculated, and as a contour of supersonic parts of the nozzle, the form of a streamline $\psi = 0.9$ was used.

Thus, the value of specific output power of laser radiation, which is possible to use as fuel benzole and toluene, and as an oxidizer and ballast gas - dry air, is ≈ 9 9.5 kJ/kg. are presented in Table 3. At use of kerosene, this value is approximately 2 kJ/kg less.

By selection of fuel, except for power efficiency, it is necessary to take into account production and toxic characteristics, and cost of components.

The application of benzole is inconvenient at low environment temperature. The freezing temperature is 5.5°C. For toluene this value is equal to -80°C, and for kerosene -38°C.

Table 3. Values of specific output power of laser, based on different fuels

Fuel	k_v, m^{-1}	w_{max}, kJ/kg	w_{out}, kJ/kg
Benzole	0.41	23.6	9.6
Toluene	0.39	23.1	9.2
Kerosene	0.32	20.8	7.4

The limit for a room vapor concentration of researched fuel has the following values: 5 mg/m^3 for benzole, 50 mg/m^3 for toluene, 300 mg/m^3 for kerosene.

From said it is possible to conclude that the application of benzole is unwise, as it slightly surpasses toluene according to power efficiency, but is significantly more toxic than toluene and is less convenient in operation.

As to selection between toluene and kerosene, the application of toluene provides specific power increase of approximately 25%. However, use of kerosene has a number of advantages. First, kerosene is approximately 3 times cheaper than toluene. Second, kerosene is fuel for airplane and engines. Therefore, at the arrangement of the laser installation, working with kerosene on a flight vehicle the additional system for fuel storage is not required.

To conclude this chapter, it is possible to note the following.

The homogeneous laser working on combustion hydrocarbon fuel products is the most reasonable for development of the mobile installation.

The specific requirements to the mobile laser installation impose limitations on using this or that fuel, oxidizer and ballast gas. The main limiting factor is the required long duration of continuous operation. The estimations have shown that summarized weights of fuel, oxidizer and ballast gas, necessary for GDL operation, is very high and is unacceptable for arrangement on a flight vehicle. At the same time, the main part of the weight of fuel components used for GDL operation mass is the weight of an oxidizer and a ballast gas. In this connection, use of free air compressed by the aviation compressor as an oxidizer and a ballast gas, is the only way to organize operation of this installation.

The comparative analysis of various fuels has shown that toluene and kerosene are the most reasonable. The application of toluene provides specific power of the installation approximately 25% higher than that of kerosene. However, use of kerosene has a number of advantages. At first, kerosene is approximately 3 times cheaper than toluene. Secondly, kerosene is the fuel for airplanes and helicopters. Therefore, arranging the kerosene operating installation on a flight vehicle, an additional system for storage fuel, is not required.

Selection of the Power Installation for GDL

Selection of a Schematic of the Power Installation

In the previous chapter devoted to the computational substantiation of the basic power characteristics, the capability of development of the mobile CO_2 - GDL with use of kerosene as fuel, and as an oxidizer and a ballast gas - atmospheric air supplied by the compressor is shown.

Besides, some GDL parameters, to help determine the required characteristics of the compressor are added.

The calculations have shown that in case of GDL, the condition Po/Pn = 20 should be realized

Where: Po - stagnation pressure of a working mix(mixture) in GDL channel;
Pn - atmospheric pressure.

In view of channel losses from the compressor to GDL inlet of the chosen compressor should have the pressure ratio not less than 22 - 23.

Temperature of inlet air in the GDL combustion chamber chosen at calculations is 720K, which corresponds to the temperature range of compressed air for compressors with $P_K > 22 \div 23$.

For an estimation of the necessary value of airflow rate through GDL, it is rational to set specific energy extraction at a known accepted power range of 100 kW. As was shown in the previous chapter, the GDL specific power with combustion of kerosene in air is 7.5 kJ/kg. However this value can be reached at dry air only. At an increase of humidity of atmospheric air, the specific power begins to decrease, and at humidity value of 100% (T = 25°C) it is 5.4 kJ/kg.

To achieve the higher specific power, it is possible in principle to include an air dryer in an air flow path, however its large overall dimensions and power consumption can result in technical and operational problems and an automatic control of dryer operation is necessary if air humidity and fuel have changed. It would be more convenient to have certain losses of specific power in humid air, having compensated losses by increased air flow.

Therefore, having accepted average specific power of 6 kJ/kg, instead of 7.5 kJ/kg we shall receive an estimated value of an air flow – 16.25 kg/s.

Thus, for an operation of the laser installation in power range of 100 kW, it is necessary to have the compressor with an air flow rate $\approx$ 16 kg/s and $P_K > 22 \div 23$.

In essence, the task of air compression at flow rate of 16.25 kg/s and pressure in access of 20 atm does not contain unsolved technological problems, since compressor engineering is widely applied in industry. The task consists of finding the optimal solution ensuring a reasonable cost of the installation, reliability of its activity and convenience in operation. Let's consider three possible schematics of power installation.

Development of Special Power Installation for the Above-Mentioned Parameters of the Compressor

In this case, the installation will consist (Figure 6) of the compressor itself, turbine for driving, gas generator for rotation of the turbine, and the fuel supply system, control and regulation units.

As there is no installation for the mentioned parameters in industry, it is necessary to develop such installation from very beginning, to conduct preproduction activity and to find a producer. For all reasons, the price of the produced installation will be high, and development will be time consuming.

Besides, short order book for such installations (single production) will initially lead to an unprofitable operation for any producer.

The Installation of Air Compressor on Aviation Engine Shaft

In general, the decision can be rather simple. On the shaft of the aviation gas-turbine engine, the compressor with the required characteristics is installed instead of the propeller.

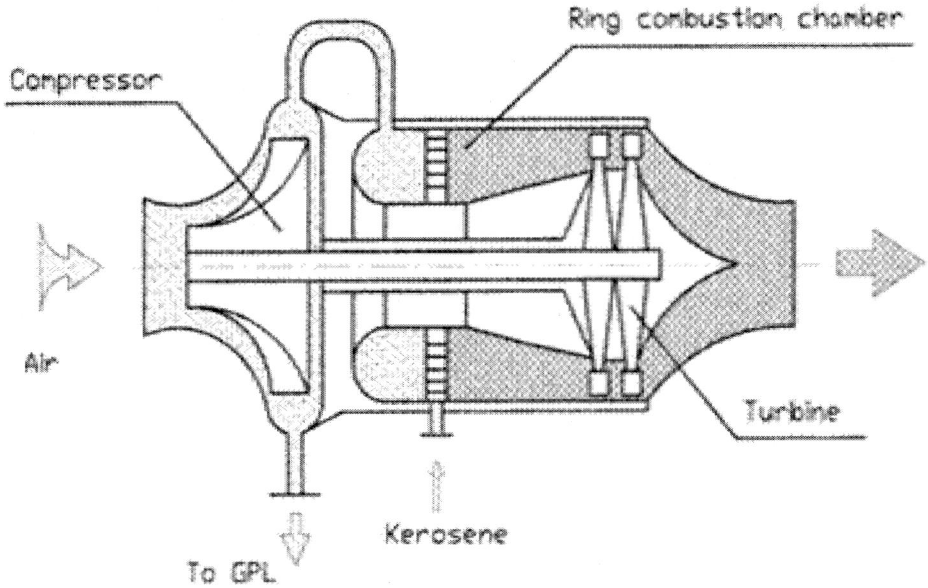

Figure 6. The new developing installation.

The calculations show that the shaft horsepower about 5.5-8 MW is necessary for compressor driving with the flow rate of 16 kg/s and Pк 23 kg/cm^2. The aviation gas-turbine engine is of middle power class, a lot of which is in series production. However, as to engineering realization, the task has a number of serious problems.

The nominal speed of the propeller shaft of existing aviation engines is in the range of 5-8 thousand rpm, and all high efficiency centrifugal compressors require operation speed of the order above. For example, industry compressor TB7-117, being far from the best in efficiency, already requires 30000 rpm.

The installation between the gas-turbine engine and compressor of the high-speed gear reducer (multiplicator), because of complexity in operation and large overall dimensions, excludes such solution as acceptable.

Technically the task is solved by the installation on the shaft of the engine of an axial multistage compressor (Figure 7). But even the principal diagram general view demonstrates the technical irrationality of such solution, it has practically two aero-engines, one of which is a production one, and the second requires a large design and manufacturing costs connected to refurbishment for realization of rigidity of the shaft, organization of turbine side exhaust, etc. Besides that, the overall dimensions and accordingly, the weight of the installation, are considerably increased.

Air Bleed from Aero-Engine Compressor Final Stage

As it is visible from the scheme (Figure 8), if it is decided to take an air bleed for GDL after a final stage of the engine compressor, practically there is no need to redesign the power installation by installing new aggregates. Moreover, the dismantling from the engine of the free running turbine operating for driving the propeller shaft for the reason of propeller nonuse, considerably simplifies a design of the installation as a whole.

This solution bribes by the simplicity, since the engines will not only be used with minimum modification, but also all control systems are kept, including the use of the control panel which is taken off from a flight vehicle.

Besides, from point of view of power, dismantling the free running turbine together with nozzle block allows lower required power of the turbine of system of the compressor - free running turbine.

The decrease of required power of the system offers the possibility to reduce the quantity of required air and kerosene in the combustion chamber of the gas-turbine engine. With constant power rate of the turbine for the drive of the compressor and of the compressor flow rate, the possibility probably exists to take off a part of the air and to direct it to the combustion chamber of the gas dynamic laser. At the same time, utilizing the characteristics of the combustion products down the main turbine (ρ, m, T), it is possible to organize an ejection of GDL gases down the supersonic diffuser for improvement of the gas dynamic laser overall performance, especially at the start.

Selection of the Aviation Gas-Turbine Engine

With the purpose of optimal selection of an aviation-engine for this task, it is necessary to formulate the initial requirements.

1. For obtaining air pressure in GDL inlet not less than 20 atm and in view of losses in a flow channel, the compressor of the engine Pк should be in the range of 22-25, not less.
2. The GDL inlet air flow rate is 16.25 kg/s.

   As for elimination of extreme engine power reduction, the air bleeding from the compressor should not exceed 15-18%, the total air flow rate through the engine should be not less than 100 kg/s.
3. The chosen aviation engine should widely be used by standard aviation operation, that enables to utilize engines, which are decommissioned from flight vehicles, having passed prescribed operational life limit are still operable for ground installation. To buy them is much cheaper than to buy new engines.
4. It is desirable to make a choice of domestic (Russian made) engines that enables in case of realization of the project to connect to activities, both engine developing company and manufacturer.

The table data of some Russian engine characteristics: Пк and G of airflow rate (kg/s) are given below.

Engine	НК-12СТ	НК-14СТ	НК-14Э	НК-16СТ	НК-17	НК-36СТ	ПС-90А	РД-33
Pк	8,8	9,5	9,5	9,68	9,68	23,12	19,6	22,0
G (kg/s)	56,0	37.1	39,0	102,0	102,0	101,4	56,0	80,0

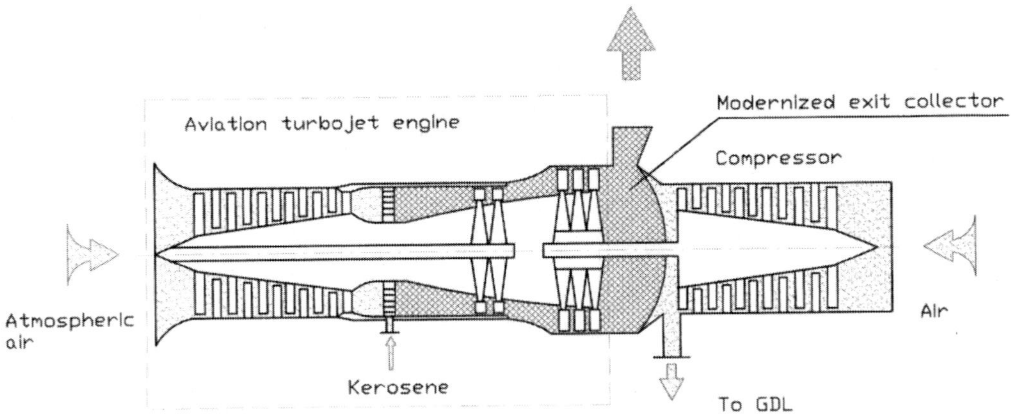

Figure 7. Addition of air compressor to jet engine.

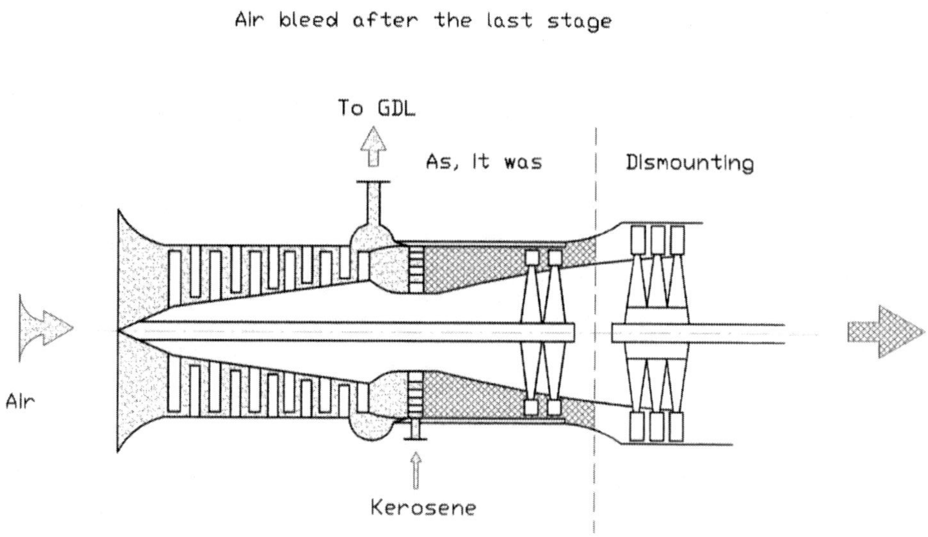

Figure 8. Air bleed after the last stage.

The analysis of the table results in a conclusion that the most reasonable for the laser installation could become the engine НК-36СТ, serially produced by Samara plant "Motorostroitel", which satisfies all requirements presented above. It is obvious that this is not the single engine, which can be put to use for this task, but it would be desirable to note the following advantages.

The high performance engine НК-36СТ of the Samara Technological complex "Motorostroitel", was developed in 1990 on the basis of an aero-engine НК-321 and is designed for the drive of the centrifugal supercharger in a structure gas pipeline pumping aggregate. The modular design of the engine facilitates transport and assembly. The engine has the remote control panel and record of long life ground operation. The general view of the engine is shown in Figure 9. In Figure 10 the design concept is shown.

Figure 9. Gas-turbine engine NK-36CT.

Gas Dynamic Assesment of the Chosen Aero-Engine as the Power Unit for Mobile GDL

In the previous chapter, the task of GDL supply with oxidizer (air) by use of the НК-36CT engine compressor was considered.

However, one more function - use of the engine exhaust gas as a working medium for ejection of a GDL exhaust system, is planed for the chosen engine.

The practice of development GDL with direct exhaust of combustion products in the atmosphere has shown that the aft diffuser of the laser produced the pressure recovery of gases up to atmospheric pressure required for starting the combustion chamber pressure increase (more than 30 atm), or decreasing exit pressure below atmospheric one. It is possible to ensure the last requirement, through installation of an ejector, for which the exhaust gases down the engine turbine are the working medium.

Gas dynamic estimation of the capability of the НК-36CT engine of GDL diffuser start support is shown below.

As it was already indicated, the basic adaptation of the НК-36CT gas-turbine engine is the elimination of design structure of the free running turbine, which allows to improve characteristics from the point of view of possible application in the laser installation, for which it is necessary to bleed air with the consumption $m_в = 16.25$ kg\sec by pressure $P \approx 22$ kg/cm^2 and temperature $T = 720$K.

Reasonable temperature for a GDL gas flow channel is $T = 1600$ K at combustion-chamber pressure of 20 kg/cm^2.

The given temperature is realized in the combustion chamber for fuel kerosene - air at mass ratio $K_к = 38.5$, that is at kerosene flow rate $m_n = 0.422$ kg/s.

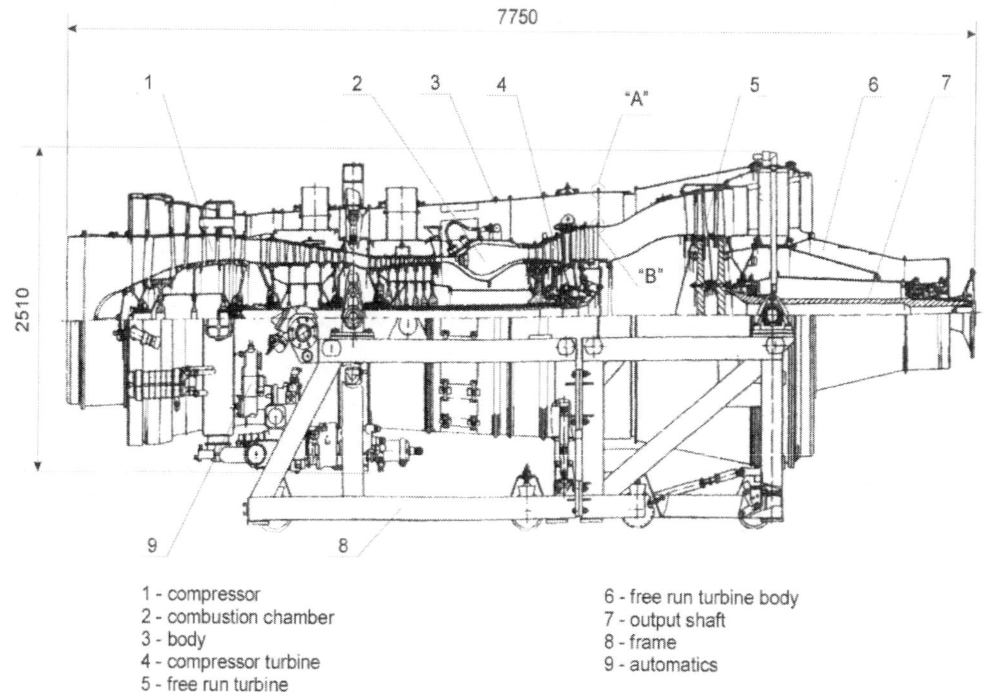

Figure 10.

In the GDL working section, it is necessary to ensure static temperature of combustion products Ta = 370 K, which is realized velocity exhaust of combustion products through supersonic nozzles with velocity Wà = 1700 m\s at a Mach number Ma = 4.46, isoentropy index κ = 1.333, pressure Pa = 0.057 kg/cm^2.

For realization in GDL flat working section shock wave free flow of the combustion products in the atmosphere is necessary to utilize the flat aft supersonic diffuser. One of major parameters of functioning of the diffuser in GDL operation conditions is the start pressure.

$$\frac{Pst.}{P\text{н}} = \frac{g(\frac{1}{\lambda a})}{g(\lambda a)};$$

The start pressure (Pst) is determined on intensity of a direct shock wave relation:

Where: Pn - pressure behind the diffuser;

$\lambda a = \frac{Wa}{a_*}$ - velocity factor equal to the ratio of exhaust velocity in working section of GDL (Wa) to the critical velocity of sound (a $*$).

The value of a velocity factor λa is connected to a Mach number by a ratio

$$\lambda a = \sqrt{\frac{X+1}{2} \cdot \frac{Ma}{\sqrt{1 + \frac{X-1}{2} Ma^2}}}$$

and then to a Mach number Ma = 4,46 corresponds $\lambda a = 2{,}32$.

The function $g = \frac{Fkr.}{Fa}$ determines the ratio of the geometrical area of critical cross-section of the diffuser to the area of exhaust gases and is according to expression

$$g(\lambda a) = \left(\frac{X+1}{2}\right)\frac{1}{X-1} \cdot \lambda a\left(1 - \frac{X-1}{X+1}\lambda a\right)\frac{1}{X-1}$$

For direct shock wave the following ratio is valid

$$\lambda a \cdot \lambda_{n.c.} = 1;$$

Where: $\lambda_{n.c.}$ - dimensionless velocity factor after front of a direct shock wave

Thus, the ratio of pressure of start to pressure behind the diffuser is equal

$$\frac{Pstor.}{Pn} = 13{,}7.$$

The wind tunnel pressure of start is usually above their steady state operating pressure up to 10-30%.

At the same time, in GDL laser conditions, when there is a gas volume involved in a working section (region of resonator mirrors) pressure of start of the diffuser can still increase the value, at which the diffuser could not be started.

Hence, it is necessary to lower exhaust pressure from the diffuser, which can be realized through an ejection of a gas jet, flowing out from the diffuser, by a gas jet flowing out after the turbine of the engine (Figure 11).

The designs of similar ejectors with maximum pressure recovery are known at high speeds of flow of gas in minimum cross-section. Such ejectors designs are characterized by smooth profile of an inlet part and with small cone angles (L) on an exit, which make value L = 6 – 10 degree.

The pressure recovery coefficient for such ejectors can reach value

$$\frac{\text{Pexit}}{\text{Pinlet}} = 0{,}955 \div 0{,}965$$ and even higher.

Thus the velocity factor value in the minimum cross-section should be

$$\lambda = 0{,}75 - 0{,}85.$$

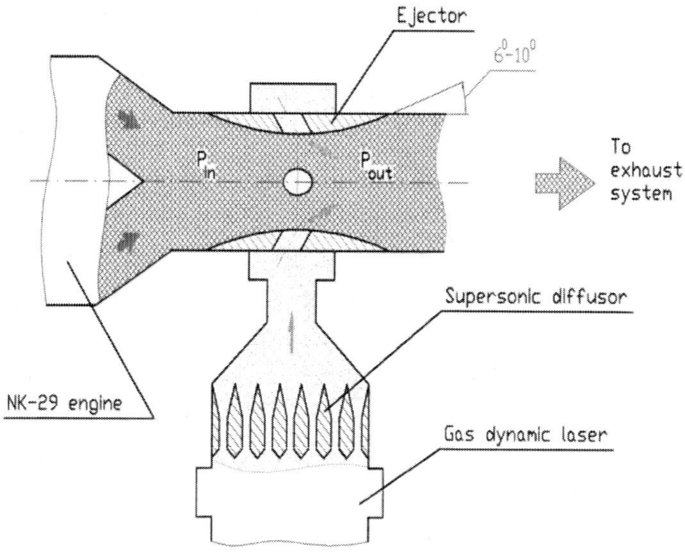

Figure 11. GDL exit gas flow ejection.

Thus, such ejector inlet pressure should be Pinlet. = 1,076 kg/cm^2, which can be provided with the gas-turbine engine without a free running turbine. (Pexit. = 1,033 kg/cm^2).

At a velocity factor $\lambda = 0{,}8$ in minimum cross-section and ejector inlet pressure Pinlet. = 1,076kg/cm^2 the static pressure in minimum cross-section becomes equal Pmin =0,73kg/cm^2, which allows to organize introduction of diffuser gas flowing out from the GDL in an area of minimum cross-section of the exhaust device of the gas-turbine engine and by that to lower pressure behind the diffuser GDL at least down to Pexit = 0,8 instead of 1,033.

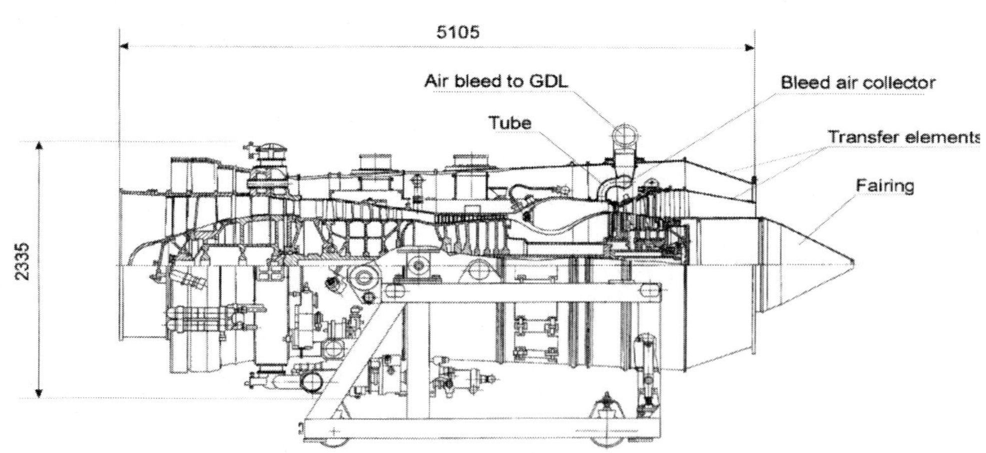

Figure 12. NK-29 engine.

Such reduction of pressure will allow a lower start pressure of the GDL down to a reasonable level of 18-20 atm.

Design Solution for GDL Power Unit on the Basis of the Gas-Turbine Engine

In view of the technical plan generated by the authors of the present technical proposal, experts of the Samara complex “Motorostroitel”, has developed on the basis of the HK-36CT gas-turbine engine the engine for a laser complex under index HK-29.

The HK-29 engine (Figure 12) represents the gas-turbine engine obtained as a result of adaptation of the engine HK-36CT, which consists of the following:

Dismantling of the Free Running Turbine with the Output Shaft

The dismantling is made through the joint "A" flange of a engine body (see Figure 6), through the joint "B" of the hot gas pipe down to the turbine of the compressor and finally the free running turbine together with a turbine casing is removed from the engine. The frame, on which the engine is mounted, consists of two not connected permanent sections, the removal of the second turbine side section does not represent any complexity.

For organization of the smooth gas flow from the modified engine, and also for fixing of the exhaust system to the engine, two conical adapters (casing) and internal fairing made of stainless steel are added.

The external adapter is terminated by a flange to be fixed with the engine ejector. The adapters and fairing were designed by “Motorostroitel”, a supplier in accordance to the aerodynamic characteristics of a gas stream and the strength requirements.

Ejector, which scheme is shown in Figure 10, is not the engine item. The ejector design is calculated and finally developed on the basis of engine, and laser complex exhaust system parameters and GDL exit characteristics (down the supersonic diffuser). Ejector design represents a pipe from high-temperature steel having symmetrical narrowing in the middle part ahead of a critical cross-section. The pipe can be manufactured from three parts with flange or welded connections.

The middle narrowing part can be turned from a thick-walled billet or from a rolled ring and has mating collars for different side collector for convenience of assembly for welding. The orifices connecting cavities of exhausts of the engine and GDL are located around the perimeter of critical cross-section. The collector is manufactured by stamping of a sheet, has the welded branch pipe and serves for uniform distribution of gas after GDL on all cross-sections of an ejector. The adaptation of the engine with removal of the free running turbine considerably reduces overall dimensions and weight of the installation. For clarity of estimation of advantage in overall dimension reduction of the modified engine, the presentations of Figures 9 and 11 are made in the same scale.

Air Bleeding to GDL

The next considered problem is the air bleeding after the compressor to GDL.

During our long search for the engineering solution for the engine adaptation, the aviation specialists have offered to realize the air bleeding from the exit of the combustion chamber second contour, not from the compressor final stage exit, i.e., from a channel of a cooling flow of the combustion chamber. This solution is connected to the fact that the combustion chamber of the gas-turbine engine is one of the most thermal stressed items of the engine, and the cooling parameters are results of long development stage. Therefore, the air bleed behind the compressor, reducing the air flow of the combustion chamber cooling flow, will change the steady state thermal status, which can result in undesirable consequences for engine operation.

The collector ring for air bleeding with a branch pipe to GDL supply was recommended to position in a zone of the exit of cooling flow of the combustion chamber, and for that the chamber and collector are connected by angular branch pipes (Figure 12). The angular branch pipes, 8 pieces in quantity, can be manufactured of a pipe or a stamped sheet from two halves and have leak tight flanges to cooperate with the envelope of a chamber cooling flow channel. The angular branch pipes are fixed by welding to the collector ring, which has the terminal flanged tap. To this tap, the branch pipe to supply air to GDL is fixed. This branch pipe passes through an outer shell of an engine body and has a "floating" attachment, excluding deformation of designs at a thermal expansion of contacting elements.

It exists from the experience of such bleed scheme, however particular design should pass a number of development tests with the purpose of experimental selection of channel pressure reduction to ensure calculated bleed air flow rate. All other elements of the engine remain without change, including all start and cutoff automatics and operation control. Thus engine HK-29, obtained as a result of adaptation of the engine HK-36CT will require minimum development, and manufacturing costs.

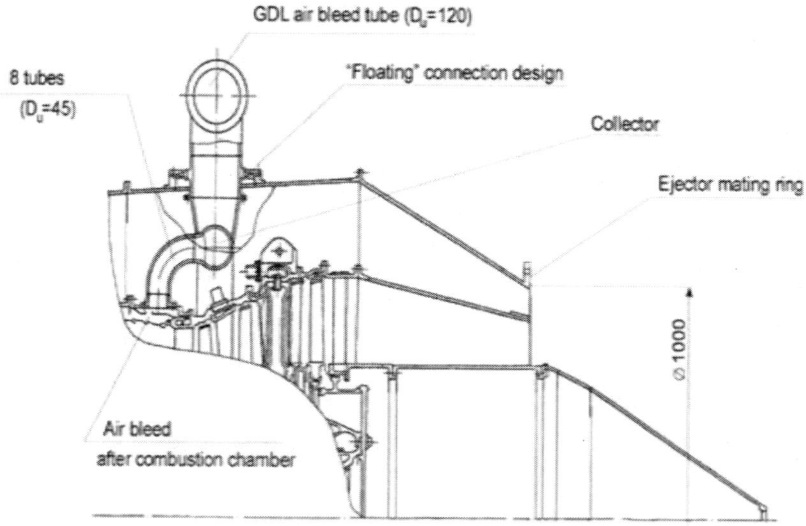

Figure 13. GDL air bleed block design.

The basic performances of the engine HK-29. All parameters of the engine are indicated for a nominal mode.

Fuel	- Kerosene TC-1
Power, MW	- 25
Air flow, kg/s	- 107,2
Compressor pressure ratio, Pk	- 23,12

CC gas temperature, K	- 1420
Gas exit temperature, K	- 698
Propellant consumption, kg/s	- 1,434
Engine weight, with frames, kg	~- 5000

The engine НК-29 is single mode, activity of the laser installation requires neither decrease, nor increase of the power, which provides reliability and life value growth.

The engine НК-29 should be manufactured by an aviation plant, irrespective of whether it is refurbished through adaptation of the engine НК-36СТ or again assembled with use of furnished engine parts from production engines.

Thus, as a result of the carried out research work, not only the optimistic results for possibility of adaptation of the gas-turbine engine for GDL gas dynamic complex were obtained, but also design activities by definition of general realization of such engine are executed in practice. It is very important that this activity is executed by the experts of aero-engines developing - producing company.

The Mobile Laser Installation

In view of all computational investigations, design studies and engineering estimations within the present paper, the provisional structure of the autonomous laser installation for clearing surfaces of reservoirs of petroleum recordings is produced. The general view of laser installation is shown in Figure 14. As it can be seen from the figure, the installation represents modular assembly. The engine has its own fixing frame, autonomous subsystems and producer - supplier. GDL also represents the complete aggregate, the developer and manufacturer of which is not connected to other suppliers. The same concerns are valid for the fuel storage system (not shown in figure). The task of the installation developer as a whole is the coordination of both these units on arrangement on a vehicle, and organization of interaction during operation. For example, relative arrangement of the engine and laser can be arbitrary, depending on overall dimensions and device of a vehicle, however in any case the fulfillment of following conditions is necessary:

- The engine and GDL should not have a rigid connection on the frame, the GDL frame should have vibration absorption plates.
- The engine and GDL should be placed as close as possible, as the unreasonable elongation of gas lines connecting them will result in additional losses because of pressure reduction.
- On fire protection control, the required fuel storage container should be placed at a distance from the engine and GDL, and is separated from them by a fire-prevention wall.
- For protection of the attendants the remote control panel should be born for limits of the installation, and input part of the engine and line of a beam are protected by special fences. The control of a scanning mirror is also carried out for limits of the installation.
- The direction of exhaust gases should not intersect structural items of the installation.

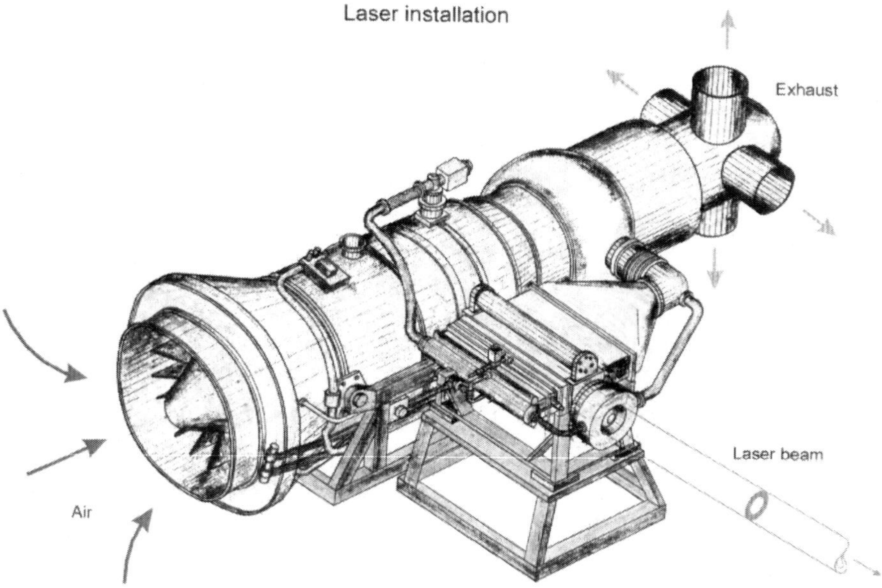

Figure 14. Laser installation.

As the development of the laser installation does not assume development of the testing site for development testing, prior to deliveries the development installation should be assembled, on which all design and operational problem solutions will be found.

CONCLUSION

The Purpose of the paper is to develop a laser technology of cleaning of large areas of seas and other water surfaces from oil film contamination. This paper also aims at development of composition scheme, technological requirements and engineering design solution of aggregates and units of vehicles, vessels or aircraft compatible mobile GDL (100–250 kW) intended to solve this important problem of environment protection. This method proposed for development is expected to complement other traditional methods, which are usually more successful at treating bulk layer oil pollution but do not match to eliminate up to 100 m oil films, the latter usually being spread over numerous square kilometers of sea and kilometers of seacoasts. Note, about one million tons of such films drift now in the World Ocean, as consequence of such disasters.

Thus, this promising approach has a fundamental science basis, from one hand, and skill and experience of missile jet specialists and their laser prototypes, from another hand. The investigations are planned to be accomplished in three basic directions:

1. theoretical and experimental research of the laser radiation action on thin oil and it's derivatives films over water surface, evaporation in three basic directions:
2. experimental optimization of laser optical scheme and operation modes in order to achieve the maximum optical and gas – dynamic efficiency.

3. engineering design, composition scheme and accompanied problems solution in order to adapt the laser module for vehicle, ship or aircraft transportation.

Results expected are as following:

- Certain recommendations on oil films elimination from large water areas, including conditions of real tanker and pipeline disaster, by means of laser radiation; development of technologies of oil and it's derivatives films burning out (mode number one) and oil films sucking up and saving of oil products for future efficient usage (mode number two) by means of high power mobile CO_2- GDL.
- Development of self-contained mobile CO_2 GDL tailored for these environmental problems solution.

These investigations may result in significant progress in the following branches of science:

a) physics of liquid inhomogeneous films, phenomena on the boundary of water-carbohydrate composition, phase transitions and chemical processes in two-phase or multi-phase compositions under high power CO_2 laser radiation and related phenomena;
b) problem of active resonators for high power GDL, different methods of laser efficiency calculation and simulation, control of temporal radiation modulation and optimization of performance, with respect to the problem of beam quality.

The author has completed the conceptual analysis of GDL installation development for disposal of petroleum films from a water surface. The basic capability of development of such installation is shown. The basic characteristic calculations of the system are made, the schematic and design solutions of basic installation are presented.

Besides that, the GDL bench has been restituted by the author of the paper and his colleagues on their own costs and some demonstration operations with scanning the beam over water surface covered with petroleum film were carried out. The experimental confirmation of film effective burning mode and mode of oil film gathering was obtained. The next part of realization phase can be the development of the installation preliminary design, in case of the customer objective and specific technical tasks, formulated by him. The preliminary design will include complete study of a design of all elements of the installation, technological works, will define all suppliers of the developing systems and sub suppliers from obtaining from them the consent on terms and conditions. In the preliminary design, schedules of all phases of installation development will be worked out: issue of the working documentation, technology modification, manufacturing of a prototype, prototype development and beginning of installations deliveries.

Thus, according to the estimation the experimental mobile CO_2 GDL installation for water surface cleaners can be designed, manufactured and tested for not longer than two – two and half years.

REFERENCES

[1] V. V. Apollonov, A.M. Prokhorov, “Universal laser for industrial, scientific and ecological use”, Proceedings of GCL/HPL Conference, St. Petersburg, p. 140, 1998.

[2] V.V. Apollonov, “Ecologically safe High power Lasers”, Proceedings of Lasers-2001 Conference, p.3, 2001.

[3] V.V. Apollonov, “High power autonomous CO2 GDL (100kW,CW/P-P modes) for new technologies development and environment protection”; Russia –NATO Int. Seminar, Moscow, 2005, Proceedings of GCL/HPL Conference, Lisbon,p.35,2008.

[4] V.V. Apollonov "New application for high power high repetition rate pulse-periodic lasers”, “Laser pulse Phenomena and Applications” Intech,p.19,2010.

[5] V.V. Apollonov, “Oil films elimination by laser”, Oboronzakaz, №17, December, p. 33, 2007.

[6] V.V. Apollonov, Yu. S. Vagin, V.V. Kijko, V.I. Kislov “High rep. rate P-P lasers”, Patent RF №2175159.

Chapter 17

LIGHTNING PROTECTION SYSTEM BASED ON HIGH POWER/ENERGY LASERS

PACS: 52.80.Mg; 52.50.Jm; 42.55.Lt; 42.62Cf.

INTRODUCTION

The feasibility of producing continuous laser sparks (CLS) with low resistance by focusing radiation from CO_2 laser with a conic mirror is demonstrated. The laser energy input per unit length required for this is experimentally found to be equal to ≈200 J/m. The possibility to efficiently control the trajectory of an electric discharge by means of CLS is demonstrated. The effect of polarity in the electric breakdown of air gaps between the CLS plasma channel and a metal rod is discovered and interpreted. The transverse structure of the CLS conductivity is investigated. The possibility of a long laser spark (LLS) with much higher resistance by focusing radiation from CO_2-laser with a spherical mirror used for protecting objects against lightning is studied. Conditions under which the electric charges from clouds can be guided reproducibly along LLS are found. Experiments reveal that the interaction between the LLS and the discharge from an electrode (lightning rod) leads to a decrease in the lifetime of the streamer corona burst, as well as to an increase in the current of the developing leader and its velocity compared to the case without the LLS.

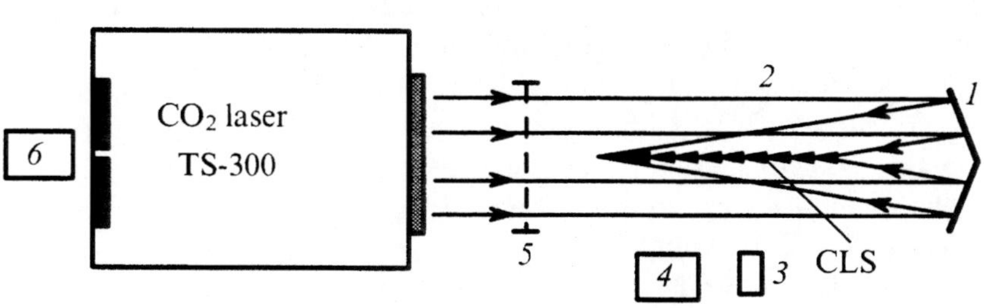

Figure 1. Schematic of the setup for CLS production: (1) comic mirror; (2) laser beam; (3) photodiode; (4) digital photographic camera; (5) transmission bolometer; (6) photodetector.

The problem of protection of terrestrial objects against lightning has not been solved satisfactorily so far. Lightning strikes on industrial objects like petrochemical complexes, nuclear power plants, ultrahigh voltage transmission lines and substations, rocket-launching devices, etc., may turn out to be extremely hazardous. The electromagnetic pulse from a close-range strike of lightning also results in a breakdown of telecommunication equipment and its operation, computer networks, and control systems. Consequently, stringent requirements are imposed at present on lightning protection measures [1, 2]. The traditional lightning protection systems being used currently are not always in a position to ensure the desired level of efficiency. This stimulates the quest for new approaches to solve this problem. Laser protection against lightning is one of the most prospective trends that are being developed actively at present [3, 4]. While using this approach, it is assumed that the lightning discharge channel being developed is guided towards the conventional rod of the metal lightning rod along the plasma channel formed as a result of the laser induced breakdown of the atmosphere. This method is based on the concept of an active lightning rod, when a laser beam can be used for "triggering" and guiding a positive ascending leader from the tip of a lightning rod to a negatively charged thunderstorm cloud. It is expected that in contrast to the traditional approach, the use of laser spark will make it possible to control efficiently the very process of protection from lightning, ensure the selectivity of lightning capture, and provide safety of tall objects and large areas. In most experimental works, the effect of the LLS length on the process of discharge development was studied only in the gaps between electrodes [3, 5 and 6], and an attempt at striking a lightning in the field tests with the help of LLS was made only in Refs [3, 7 and 14]. Two cases of simultaneous occurrence of a strike on the lightning rod and triggering of a laser were registered in [7], but the lightning channel did not coincide with the optical breakdown channel. On the other hand, it is known from the experiments on initiation of discharge in the gap between electrodes that LLS completely determines the development of the leader channel and its position in space [5].

However, the conditions under which a long laser spark can initiate an ascending leader or trap a lightning strike are not clear, and the field tests of laser protection against lightning carried out in Refs [3, 7 and 14] could not provide an answer to these questions. One of the most promising methods for producing a plasma channel or controlling the trajectory of an electric discharge is the exposure of gas to laser radiation. Of all kinds of laser action, an optical breakdown leads to the longest and most intense gas perturbation. The length of optical breakdown channel, under real atmospheric conditions can amount to tens of meters and over [18, 19], which makes it possible to produce guided electric discharges several meters long [3, 8]. In Ref. [15], the length of a guided discharge was equal to 16 m for a 3-MV voltage across the discharge gap and a 460-J output energy of a pulsed CO_2 laser used to obtain the LLS. However, an LLS produced by spherical optics consists of separate breakdown nuclei located at random in the caustic of the focusing element, i.e., it is immanently discontinuous. For this reason it has not been possible to attain good reproducibility in attempts to control the trajectory of very long discharges with the use of LLS. This is also evidenced by the results of the tests of a laser lightning protection system under real thunderous conditions in Japan, which failed because two events of time coincidence a thunderbolt to the lightning rod with the triggering of a laser were observed, but the trajectories of the lightning channel and the LLS did not coincide [22]. A continuous optical breakdown channel devoid of gaps, or a continuous laser induced spark (CLS), can be obtained by focusing laser radiation with a conic lens (an axicon). A =1-m long CLS was experimentally obtained [21] by focusing Nd-laser radiation with an axicon. The electrical characteristics of a 9-12 cm long CLS produced by an Nd laser emitting 40-ns pulses were investigated in Ref. [10]. For output laser energy of 100 J, the resistance of the spark under investigation was 0.5-1 Ohm. On applying a voltage of 1-25 V across the spark, the electric current reportedly started flowing immediately after the optical breakdown. The results of Ref. [10] demonstrate that the efficient control of the trajectory of electric discharges is possible using CLSs produced by focusing laser radiation with conic optics. However, considerably longer CLSs may be needed for practical applications. For instance, estimates made in Ref. [22] show that the plasma channel length in real lightning protection systems should be equal to -20 m., increasing the wavelength of laser radiation results in a lowering of the optical breakdown threshold, which should permit increasing the CLS length – a parameter critical for practical applications [23]. In this paper, we investigate the feasibility of producing a CLS with the use of a CO_2 laser and also the feasibility of its application for guiding electric discharges and transferring the electric potential. The schematic diagram of the setup for studying CLSs is shown in Figure 1. The radiation of an electric-discharge TS-300 CO_2 laser [24] with a maximum output energy of 300 J and the radiation divergence at half-maximum of -3.5x10^{-4} rad was focused with conic aluminum mirrors (1) with a cone angle g = 177.8 or 179.16°. The angle of incidence of the laser beam (2) on the focusing mirror did not exceed 5°. The radiation of a laser spark in the visible range was recorded with a fast photodiode (3) and was also photographed with an Olympus C-3030 digital camera (4). The laser beam with a cross section in the form of a frame with external and internal dimensions of 10x10 cm and 3.3x3.3 cm, respectively, was formed by an unstable telescopic resonator with a magnification M = 3. The out-put laser energy was recorded with a transmission bolometer (5) in every shot; the shot-to-shot energy spread was within 5%. The output laser energy was varied during experiments by varying the pressure of the working laser medium and the voltage of the pump source, because introducing plastic film or metal mesh filters into the laser beam resulted in beam wave front distortions and deterioration of CLS quality.

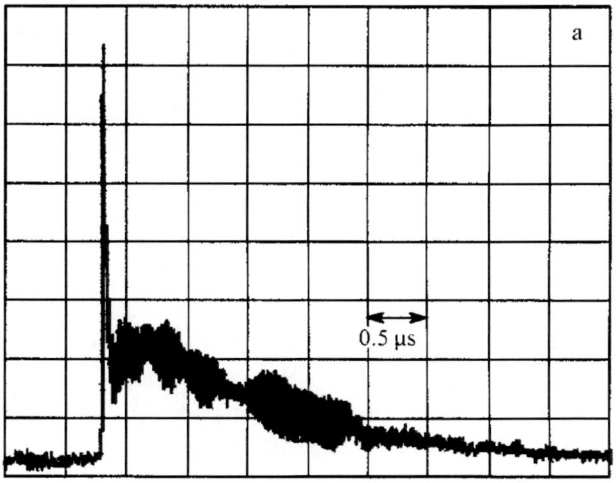

Figure 2 Continued.

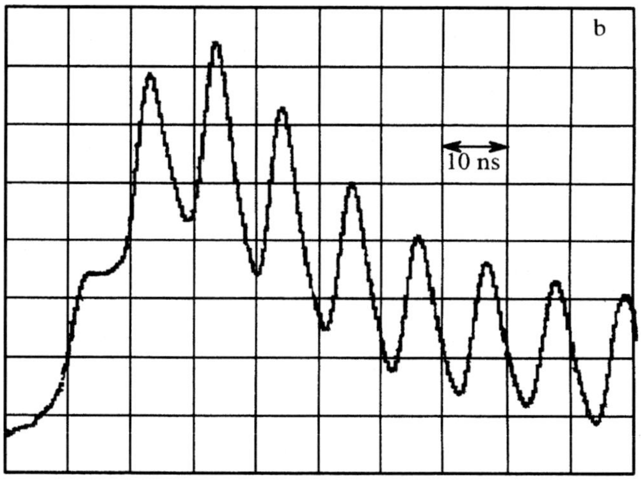

Figure 2 Oscilloscope traces of the pulse of CO_2-laser radiation.

The shape of output laser pulses was monitored with an FP-3 photon drag detector (6) through the alignment opening in the rear resonator mirror. Typical oscilloscope traces of an output laser pulse are given in Figure 2. One can see from Figure 2a that the shape of the output pulse is typical for electric discharge CO_2 lasers – a short (40 ns) primary peak and a long (3 μs) tail. Due to spontaneous mode-locking, the pulse was modulated with short (3 ns) peaks spaced at 10-ns intervals (see Figure 2b). The electrical CLS characteristics were investigated employing the circuit shown in Figure 3a. The ends of two metal rods 2 mm in diameter were introduced into the CLS plasma. The rods were connected to oppositely (relative to the common ground) charged capacitors C with capacitances of 0.25 μF each and were variously spaced in different experiments. The highest capacitor charging voltage was 50 kV, so that the potential difference across the rods could be as high as 100 kV. When studying the electric breakdown of the air between the metal rod and the CLS plasma column, one or both rods were shifted perpendicular to the CLS axis through a distance required in the experiment. The polarity of the rod moved aside could be altered. The circuit for studying the electric field distortion and potential transfer with the aid of LLS is shown in Figure 3b. In these experiments, one of the rods dipped into the plasma was connected to a resistive voltage divider (R_1 =5 kΩ, R_2 = 60 Ω). The other rod remained connected to the capacitor, whose charging voltage did not exceed 600 V, allowing us to avoid the development of ionization in the electric field. The electric field perturbation and the potential transfer during the production of a CLS were also investigated employing a capacitive voltage divider. The high-voltage arm of the divider was the capacitance between the CLS plasma channel, or the electric discharge channel, and the metal plate mounted on the divider. This capacitance could be varied by varying the plate area and the distance between the plate and the plasma channel. A ceramic capacitor, whose second plate was connected to the grounded body of the divider, served as the low-voltage arm the voltage dividers were calibrated directly at the measurement site by applying voltage drops of different amplitude across them from a capacitor C (see Figure 3b). When calibrating the capacitive divider, the CLS was simulated by metal rods of different diameters. As in the setup of Figure 1, the instants of spark initiation and the electric discharge development in it during the study of electrical CLS characteristics were recorded with a fast photodiode. The signals from the photodetectors and voltage dividers were recorded employing a Tektronix TDS 220 oscilloscope with a pass band of 100 MHz. The photograph of a CLS produced by focusing the CO_2-laser radiation with a conic mirror with a cone angle $\gamma = 177.8°$ is shown in Figure 4a. One can see that the laser spark looks like a continuous column consisting of bright cone-shaped elements. The breaks characteristic of the arks produced by beam focusing with spherical optics are absent in this case.

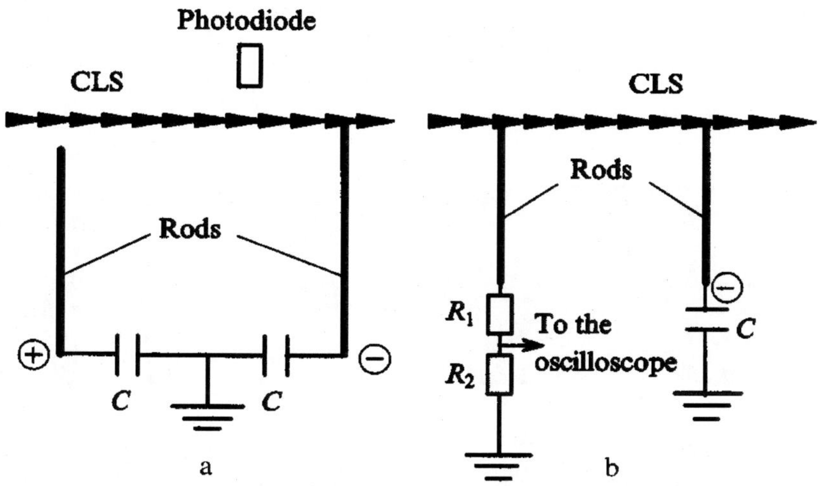

Figure 3. Circuits for study of the electrical characteristics of a CLS.

This pattern is observed for output laser energy of 180-300 J. The CLS length, determined by the cone angle and the beam dimensions in the cross section of the cone, was equal to -90 cm. The CLS onset is displaced from the mirror surface because the radiation is absent in the beam center. When the output laser energy was less than 150 J, at the ends of the spark, where the input of laser energy is lower than in the central region, visible breaks appeared in accordance with the energy distribution over the beam section. The photograph of a CLS fragment is shown in Figure 4b, in which the longitudinal spark structure in the form of bright cone-shaped elements is clearly visible. The same structure was observed when the CLS was produced by Nd-laser radiation [10, 21], but the structure period and the transverse dimension of the CLS were significantly smaller than in the case under study due to the different emission wavelengths of Nd and CO_2 lasers. For comparison, Figure 4c shows the photograph of an LLS obtained by using the mirror with a cone angle γ=179.16° for an output CO_2-laser energy of 300 J. One can see that the increase in cone angle, as expected, led to the increase in the spark length up to -3 m. However, the spark was no longer continuous, because the laser energy inputted per unit length of the spark was not high enough. The transverse dimension of the plasma nuclei in the spark also increased appreciably (by a factor of $\approx$ 1.5). However, even for an output laser energy insufficiently high to produce a continuous channel, the radiation focusing with a conic mirror makes it possible to attain a higher density of plasma nuclei per unit length of the spark compared to that obtained by focusing with spherical optics (the spark lengths being equal) because of the more uniform distribution of laser radiation energy over the spark length. Estimates made on the basis of photographs show that to produce an almost continuous spark structure by focusing the CO_2-laser beam with a conic mirror, a linear laser energy input of ≈200 J/m is required. Note that these estimates are valid only for the shape of output laser pulses shown in Figure 2. The use of techniques for raising the peak output power, for instance, by injecting a nanosecond pulse into the resonator [27], will reduce the output laser energy required for the CLS production. The electrical properties of the laser spark were investigated by focusing radiation with a mirror with a cone angle γ =177.8° (the total spark length was ≈ 90 cm). These studies confirmed that these sparks could be efficiently applied for electric discharge guidance. Even for an output laser energy of 140 J, when noticeable discontinuities were observed at the spark ends and the length of the continuous part did not exceed 50 cm, an electric breakdown was observed in the spark between the rods immersed into the spark plasma and spaced at $d \approx$ 80 cm for a minimal average electric intensity $E_{min} \approx$ 77 V/cm.

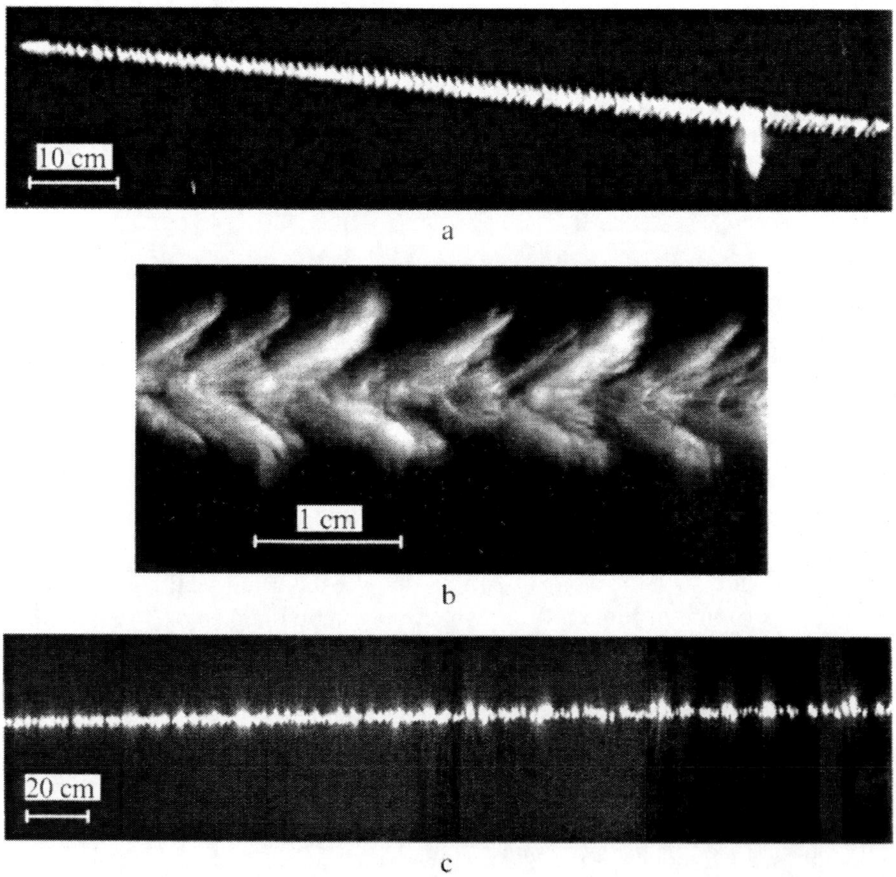

Figure 4. Photographs of an LLS produced by CO_2-laser radiation focusing with a conic mirror with a cone angle γ =177.8 (a, b) and 179.16° (c).

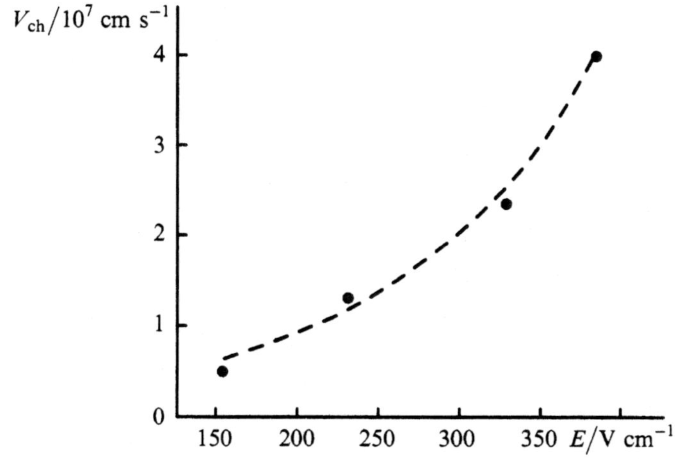

Figure 5. Velocity of propagation of the conducting channel V_{ch} as a function of the electric field intensity *E*.

Figure 5 shows the average propagation velocity *Vch* of the conducting channel of laser-guided discharge along the spark as a function of the average electric intensity in the gap *E* for *d* = 80 cm and an output laser energy of 140 J. The velocity *Vch* was estimated from the time interval between the signals from the photodiode corresponding to the instants of laser spark initiation and the appearance of the guided discharge. One can see from Figure 5 that for *E* = 385 V/cm, the channel velocity *Vch* = 4 x10 7 cm/s. The experiments on the electric breakdown of the air gaps between the CLS column and the metal rod were performed using the circuit shown in Figure 3a for a 100-kV voltage across the rods, an output laser energy of 270 J, and a horizontal rod separation d = 45 cm. For a positive polarity of the rod displaced relative to the CLS, the longest vertical distance H between the rod upper end and the CLS plasma column at which the electric breakdown occurred amounted to 13 cm.

The photograph of a laser guided discharge in this case is shown in Figure 6. Figure 7 shows the oscilloscope trace of signals from the photodiode recorded under the same conditions. The first peak in the oscilloscope trace reflects the CLS production, while the other, with a – 100 μs delay, corresponds to the current of the guided discharge. Under the same experimental conditions and a negative polarity of the rod, the longest distance H at which the electric breakdown of the air gap occurred was only 6 cm. Therefore, upon the breakdown of the air gap between the CLS plasma column and the rod, we observed a distinct polarity effect indicating that the real transverse dimension of the CLS is significantly larger than that demonstrated by the photographs (1 cm). Indeed, when the CLS was replaced by metal rods of different diameters, the polarity effect in the air gap breakdown comparable to the polarity effect observed in the CLS-guided discharge experiment manifested itself only for a rod diameter of over 4 cm. Therefore, our experiments showed that the CLS produced, by focusing the CO_2-laser radiation with a conic mirror behaves similarly to a conducting rod, its conductivity, as follows from the oscilloscope trace in Figure 7, persisting for a rather long period of time. The CLS conductivity and the feasibility of transferring the electric potential via the CLS were investigated using the circuit shown in Figure 3b for an output CO_2-laser energy of 150 J. Under these conditions, only the central - 65-cm long part of the spark was continuous. The distance d between the rods immersed into the CLS plasma was varied between 10 and 60 cm. The position of the rod under potential remained invariable in the experiment, and only the rod connected to the resistive voltage divider was moved. The polarity of the rod under potential was negative.

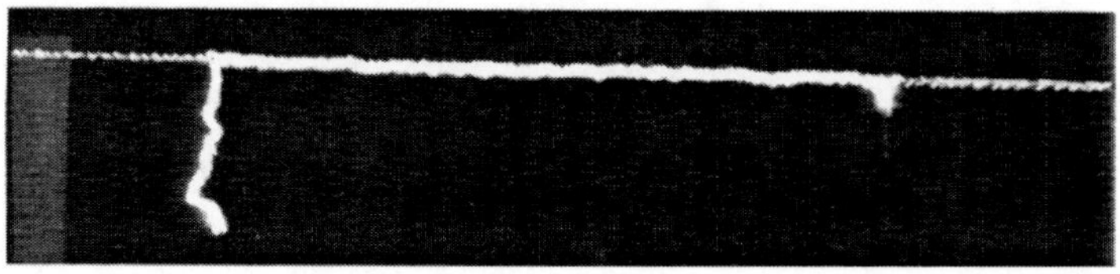

Figure 6. Photograph of a CLS-guided electric discharge.

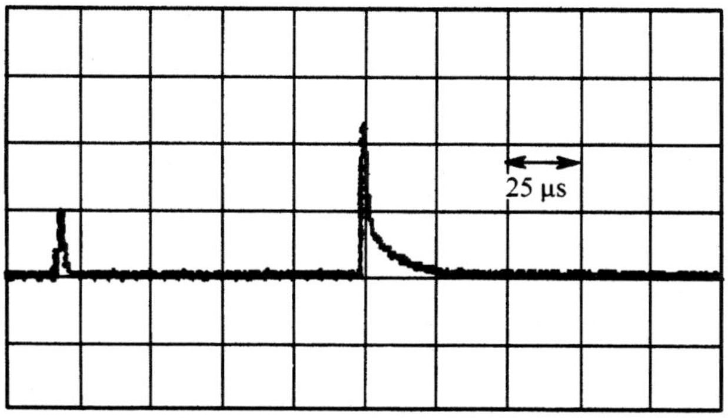

Figure 7. Oscilloscope trace of the signal from the photodiode: the CLS radiation (the first peak) and the radiation of CLS-guided electric discharge (the second peak).

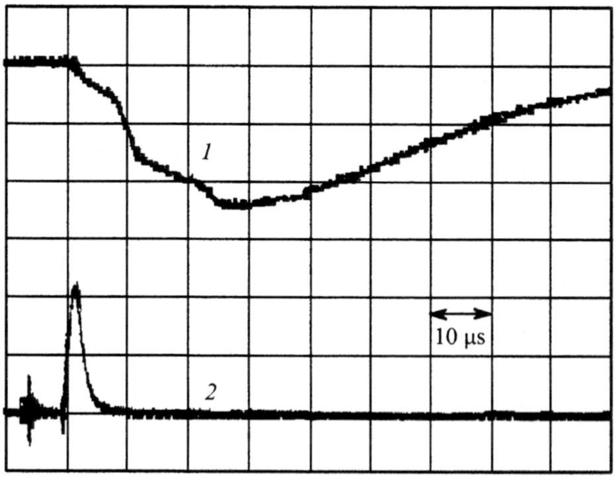

Figure 8. Oscilloscope traces of the signals from the voltage divider at the spark axis (1) and from the photodiode (2).

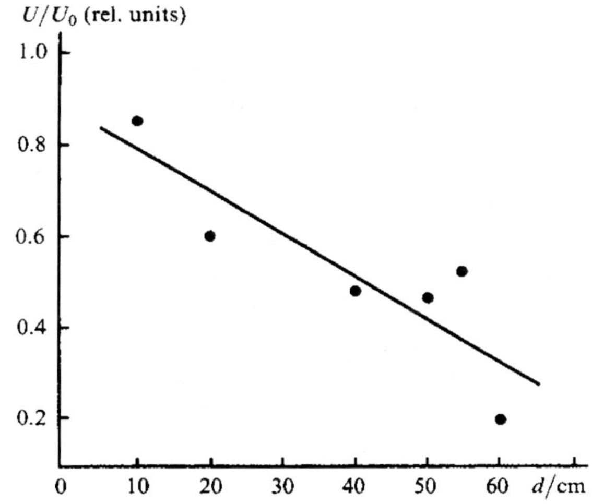

Figure 9. Dependence of U/U_0 on the distance *d* between the rods dipped into the CLS plasma.

Figure 8 shows the oscilloscope traces of the signals from the voltage divider and the photodiode (recording the CLS plasma emission) obtained for d = 10 cm. One can see that these signals commence simultaneously, i.e., the CLS is conducting since the instant of its initiation. The leading edge of the voltage pulse at the divider has a characteristic step shape, the regions of rapid voltage build-up being replaced with the regions of slow voltage growth. The voltage rise time Tf and, accordingly, the CLS plasma conductivity rise time are independent of the rod separation and are equal to -28 μs. The large duration *tf* points to an important role of gas dynamic processes in the production of CLS conductivity.

Figure 9 shows the ratio $U=U_0$ (where U is the voltage amplitude recorded by the resistive divider and U_0 is the charging voltage of the rod under potential) as a function of the distance d between the metal rods immersed into the CLS plasma. This dependence reflects the CLS-assisted potential transfer. The scatter of points may be due to an inaccurate mounting of the movable rod in the plasma channel. This dependence allows us to estimate, taking into account that the resistance of the resistive divider is Rd =5.06 kΩ, the resistance Rs of the CLS produced with a conic mirror: Rs = 988 Ω for d = 10 cm and Rs = 23:9 kΩ for d = 60 cm (the resistance per unit length is 100 -400 Ω/cm). This result confirms that the CLS conductivity is high. Note that a CLS produced by focusing a laser beam with a spherical mirror is not conducting at all due to the immanent structural discontinuities between the plasma nuclei. In the experiment being outlined, for $d > 70$ cm no signal from the resistive divider was detected as well. This indicates that the spark is non-conducting due to the appearance of breaks in its structure (recall that the experiment was performed for output laser energy of 150 J). However, the capacitive divider produced signals as d was increased up to 90 cm, which confirms the existence of significant electric field distortions in the gap induced by the spark under study, despite the discontinuities in the spark structure. This is consistent with the results of measurements of the breakdown fields along the spark considered above. Of immediate interest was the investigation of electrical conduction of the CLS in its cross section in connection with the discovery of polarity effect in the electric breakdown of CLS – rod air gaps. For this purpose, the rod connected to the resistive voltage divider was displaced perpendicular to the CLS axis. This experiment was conducted for a separation $d = 10$ cm and an output CO_2-laser energy of 270 J.

Figure 10 shows the voltage amplitude U recorded by the voltage divider as a function of the distance r between the upper end of the measuring rod and the CLS axis. One can see that the maximum distance r at which the signal from the divider is still recorded is equal to -2 cm. This is consistent with the results of the experiment on the modeling of polarity effect when the CLS was simulated by metal rods of different diameters. Figure 11 shows the oscilloscope traces of the signals from the resistive divider and the photodiode recorded for $r = 1.5$ cm.

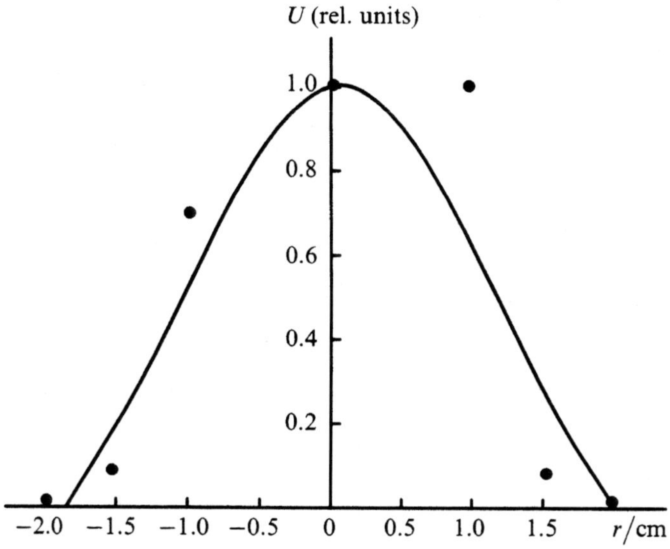

Figure 10. Voltage U across the divider as a function of the distance r between the CLS axis and the tip of the rod.

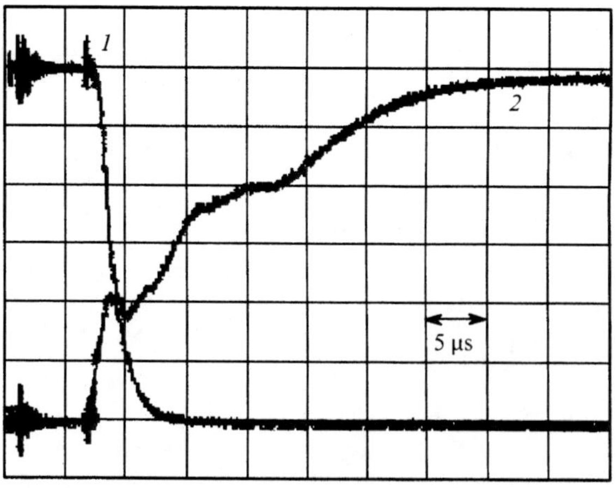

Figure 11. Oscilloscope traces of the signals from the voltage divider (1) and the photodiode (2) for $r = 1.5$ cm.

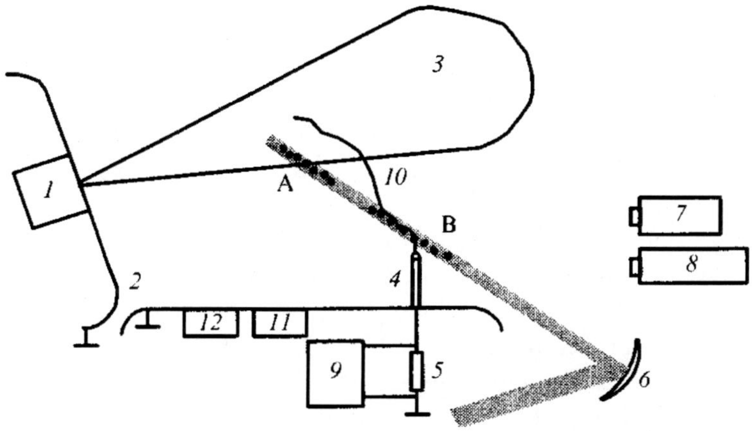

Figure 12. Schematic of the experimental setup: (1) charged aqueous aerosol generator; (2) earthed surface; (3) charged aerosol cloud; (4) electrode rod; (5) current shunt; (6) spherical mirror; (7) digital camera; (8) video camera; (9) current measuring instrument; (10) channel discharge; (11) string sensor for electric field strength; (12) dynamic aerial; (A and B) two positions of a laser spark.

Note that the voltage rise time $tf = 3$ μs in Figure 11 is substantially (about an order of magnitude) shorter than the corresponding time in the oscilloscope trace of the voltage signal shown in Figure 8, and the voltage drop from the peak value in Figure 11 occurs considerably faster than in Figure 8. Note that the duration tf in Figure 11 is approximately equal to the duration of the initial period of rapid voltage build-up in Figure 8. Therefore, we can assume that the initial CLS conductivity, both on the axis and in the peripheral regions of the CLS (in the cross section), is due to the photoionization of air by the radiation of initial nuclei of optical breakdown. The ionization corona around a spark produced by focusing a laser beam with a short-focus lens was earlier observed in Ref. [26]. However, we believe that this interesting effect invites further investigation.

Since field experiments are quite complex and expensive, it seems expedient to continue laboratory-scale studies of lightning protection systems. However, such studies must be carried out under conditions of the most appropriate simulation of the actual thunderstorm situation. In this work, we investigate experimentally as well the interaction of a laser spark with discharges induced by an artificial cloud of charged aqueous aerosol. Figure 12 shows the experimental setup consisting of two main parts: a charged aerosol generator (28) and a 50-J pulsed CO_2 laser [29]. The charged aerosol cloud (CAC) is produced by the condensation-type charged aqueous aerosol generator (1) in which a subsonic convergent nozzle with a diameter 6 mm of the outlet aperture is used. For a working pressure of 3-8 atm in the vapor generator, the velocity of outflow of the turbulent vapor jet into the atmosphere is 400-430 m/s. The results of measurement of the outflow velocity in various cross sections of a charged turbulent flow are given in Ref. [28]. The charging unit of the charged aerosol generator ensures a loss current in the range 80-150 mA. The positively or negatively CAC (3) having a volume of several cubic meters is formed at a height of 1-3 m above an earthed metal surface (2). The CAC voltage may be as high as 1.5 MV. An electric field of strength up to 12 kV/cm at the earthed surface and up to 22 kV/cm near the CAC boundaries is produced in the gap between CAC and the earthed surface [30, 31]. When the experiments were carried out, an electrode rod (4) with a spherical or conical tip was placed on the earthed surface under the CAC. The height of the electrode could be varied from 10 to 58 cm, and the radius of curvature of its tip varied from 2 to 20 mm. The CO_2 laser beam was directed in such a way that a long laser spark was formed in the gap between the CAC and the rod after reflection at the concave mirror (6) with radius of curvature 3.6 m. The spark was formed in the CAC near its boundaries (version A) or near the tip of the rod (version B). The laser spark had a length up to 50 cm. The angle between the direction of propagation of the spark and the horizontal surface could be varied from 40 to 70°. The laser spark had a discontinuous structure (up to 15 globular plasma formations along the laser beam), which is characteristic of long sparks obtained by focusing laser radiation with the help of spherical optics. The measuring part of the experimental complex in this case included low-inductance current shunts (5) (10 and 4Ω) for measuring the discharge current in the gap between the CAC and the electrode rod. The signal from the shunts was supplied to the oscillograph (9) (Tektronix TDS 220). A digital camera (7) and a video camera (8) recorded the optical pattern of a long laser spark and channel discharges (10) in this gap. The electric field produced by CAC was monitored by a string-type sensor (11) mounted on the earthed surface (2). A dynamic aerial (12) whose signal was fed to a storage oscilloscope C8-17 tracked the variation of charge in the cloud due to discharge phenomena. In this work, we present the experimental results obtained for a negative CAC, since this situation is the closest to real thunderstorm patterns in our climatic conditions. The experiment was carried out as follows. The charge of the CAC was increased to a value for which the electric field formed by it in the gap between the CAC and the rod was sufficient for creating ascending discharges from the tip of the rod. The laser was triggered manually, and a long laser spark was produced in the gap at arbitrary instants relative to the discharge processes occurring between the cloud and the rod on the surface. Series of experiments were carried out for two positions of the laser spark in the discharge gap (A and B, see Figure 12) and for two heights of the rod above the surface (58 and 29 cm). The pressure, temperature and relative humidity of the air were 740 mm Hg, 18°C and 70% respectively.

Figure 13. Photograph of a discharge with a laser spark in the gap.

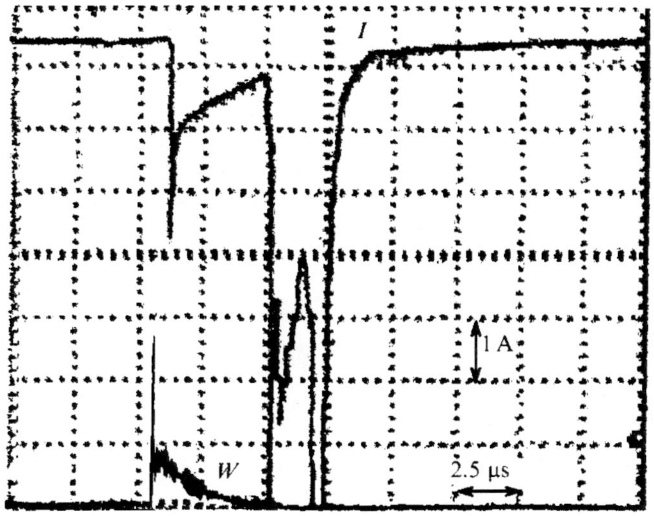

Figure 14. Oscillograms of the discharge current *I* in the presence of a laser spark (upper beam) and the shape of the laser pulse *W* (lower beam).

During the spark formation at the tip of the electrode rod (version B, height of the rod 58 cm, electric field strength of CAC 12-16 kV/cm near the tip of the rod), a strong correlation was revealed between the emergence of the laser spark and the development of the main discharge between the cloud and the rod at the surface. In this case, a part of the leader channel of the main discharge in the vicinity of the tip of the rod developed along a segment of the laser spark (Figure 13). It is significant that the leader discharge is guided not along the entire spark, but only along a part of the spark, which is not characterized by the highest density of plasma formations per unit length. The spark segment in the vicinity of the CAC boundaries does not participate in this process.

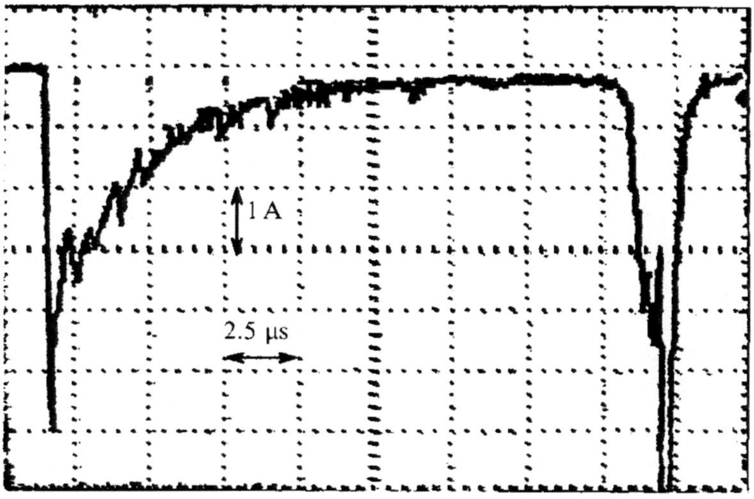

Figure 15. Oscillogramm of the discharge current *I* in the absence of a laser spark.

Table 1. Leader discharge parameters with and without a laser spark in the gap

Measured parameters	Mean values of measured parameters and their spread	
	with a laser spark	without a laser spark
Duration T_{cor} of the streamer corona burst before the emergence of the leader/μs	2.37 1.22–5.88	7.2 4.2–11.2
Maximum streamer corona current I_{cor}/A	3.37 1.88–5.0	3.95 2.46–5.0
Charge Q_{cor} of the corona burst/μC	5.0 2.07–15.8	6.7 2.8–14.8
Mean leader current I_{lead}/A	1.52 0.93–3.07	0.11 0.08–0.15
Duration T_{lead} of the leader current stage/ms	1.25 0.68–2.07	13.0 5.6–22.9
Leader charge Q_{lead}/μC	1.8 0.84–3.68	1.68 0.4–2.6
Amplitude I_m of the main discharge current/A	5.0 4.4–5.4	4.9 4.2–5.3
Duration T_m of the main discharge stage/μs	1.05 0.75–1.61	3.4 2.23–6.25
Charge Q_n of the aerosol cloud neutralized at the main discharge stage/μC	3.64 2.38–6.05	5.76 2.49–8.21
Q_n/Q_{lead}	2.28 0.98–3.79	6.7 1.1–22.6

Figure 14 shows a typical oscillogram of the discharge current passing through the rod in the case of "capture" of its channel by a laser spark. The oscillogram shows all the stages of the discharge evolution, including the streamer corona burst about 2 μs after the optical breakdown (which is transformed into a leader within about 2-4 μs), and the main discharge stage of duration -1μs following the leader after 1-3μs. If the height of the rod is reduced to 29 cm (the spark is also formed in the region B, the electric field strength near the tip of the rod is 10-14 kV/cm), only solitary acts of capture of the discharge channel by the laser spark are recorded. In this case, the duration of all discharge stages was much longer than on the oscillogram screen even in the case of capture. If the laser spark was in the region A, no connection between its emergence and the formation and evolution of the discharge was observed for all the heights of the electrode rod in spite of a high electric field strength (10-16 kV/cm) in this region (a total of 20 laser shots was carried out for this purpose). To determine the role of the laser spark in the evolution of the discharge, we carried out a series of experiments for studying the discharges in the gap between the CAC and the rod without the laser spark. Figure 15 shows a typical oscillogram of the discharge current passing through the rod in this case. The oscillogram clearly shows all the stages of the discharge evolution, including the corona burst, leader stage and the main stage. However, the duration of these stages is much longer than in the case when a laser spark is present in the gap (see Figure 14). A comparison of the discharge characteristics in the gap between the CAC and the rod, obtained with/without a laser spark in the gap, and an analysis of the conditions of their evolution show that the laser spark formed at the tip of an earthed electrode does not facilitate the emergence of scending leader discharges if the electric field of the cloud is weaker than that required for the development of a discharge in the absence of a spark. A fairly high probability of discharge channel capture by a laser spark is observed only for large values of the cloud charge, when the electric field of the CAC is in a position to initiate an ascending leader even in the absence of a spark. Consequently, the emergence of a leader with or without a laser spark occurs for close values of the external electric field produced by CAC. However, a long laser spark formed in the vicinity of an earthed electrode may accelerate the process of formation of a stable leader and its propagation towards the CAC.

Table 1 contains the results of processing of the current oscillograms for discharges "captured" by the laser spark, as well as the discharges observed without a laser spark. It can be seen that the formation of a laser spark near the tip of the electrode rod leads to a manifold decrease in the duration T_{cor} of the streamer corona burst before its transformation into a leader discharge, while the maximum current I_{cor} of the streamer corona and the charge Q_{cor} transferred by it decreased by just about 30%. On the whole, similar relations hold between the duration of the streamer corona burst and the transferred charge for discharges with and without a laser spark. It can be assumed that the presence of a laser spark near the tip of the rod in the streamer corona region accelerates the process of its transformation into the first leader segment. The velocity of propagation of the leader increases by almost an order of magnitude under the influence of a laser spark, thus decreasing the duration T_{lead} of the leader stage of discharge from 13 to 1.25 μs. In turn, the average leader current in the presence of a laser spark is an order of magnitude higher than without the spark (more than 1.5 A on the average as against 0.11 A). As a result, the total charge of the leader is found to be almost identical in both cases. Thus, the laser spark accelerates the propagation of a leader discharge if it traps the emerging leader. The laser spark may increase the average velocity of the leader manifold. This is in accord with the results obtained in Pefs [5, 6] according to which the presence of LLS spark in the discharge gap increases the velocity of propagation of the leader by more than an order of magnitude in comparison with its velocity in the unperturbed gap. It is significant that the gap is covered the more rapidly, the larger the segments of the laser spark coinciding with the leader channel. This effect is manifested clearly in our experiments also upon an increase in the length of the discharge gap between the CAC and the rod due to a decrease in the height of the rod (from 58 to 29 cm) on the earthed surface. In spite of the development of the ascending leader over the laser spark region, this led to an increase in the average time T_{lead} of its propagation from 1.25 to 5.7 μs, since the total length of the gap increases and the part covered by the laser spark decreases. The parameters of the main discharge also differ significantly in the presence of a spark in the gap and without it. Although the current amplitudes of the main discharge in both cases are 4-5 A (see Table 1), the duration T_m of the main discharge stage and the charge Q_n neutralized in this time decrease to less than half as a result of "capture" of the discharge. In both cases, a nearly identical growth of the main stage duration Tm is observed upon an increase in the time T_{lead} of propagation of the leader (Figure 16). In spite of the similarity in the increase in Tm and the increase in T_{lead}, the dependences of the charge Q_n neutralized at the main discharge stage on the leader charge Q_{lead} are of different types for charges with a spark and without it. In the absence of a spark, a tendency towards a decrease in the value of Q_n upon an increase in the leader charge Q_{lead} is manifested clearly. On the contrary, a weak increase in the value of Q_n upon an increase in the leader charge Q_{lead} is observed in the presence of a spark. It must be noted that the fraction of the charge left by the leader in the total charge neutralized at the main discharge stage ($Q_n = Q_{lead}$) is much smaller for discharges without a laser spark in the gap than in the presence of the spark (on the average, $Q_n = Q_{lead} = 6.7$ in the presence of the spark and 2.28 without it), especially for small values of the leader charge. For higher leader charges, only the charge left by the preceding leader is virtually neutralized under all conditions at the main discharge stage; i.e., $Q_n = Q_{lead} \rightarrow 1$.

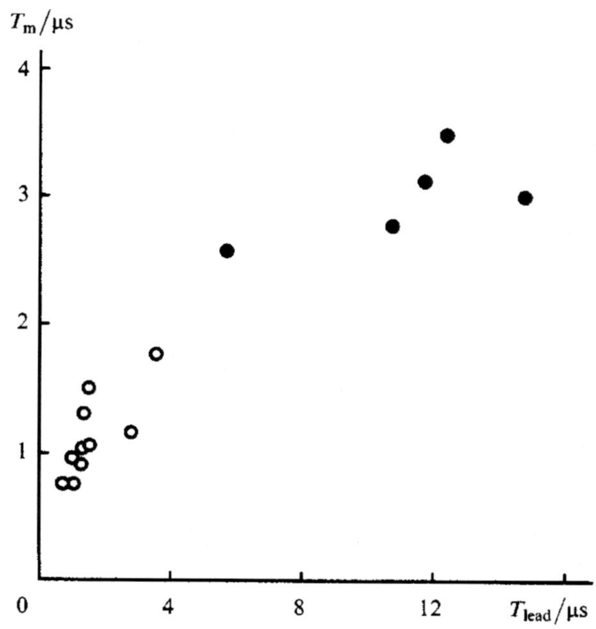

Figure 16. Duration T_m of the main discharge stage as a function of the duration T_{lead} of the leader current in the presence (○) and absence (●) of a laser spark.

CONCLUSION

Experimental studies have shown that, in spite of its discontinuous structure, a LLS created in the vicinity of an earthed electrode may capture the channel of leader discharge developing in the gap between the electrode and a CAC. The capture takes place for a quite high electric field strength at the electrode, its value being close to that of the field required for the emergence of an ascending positive leader in the absence of a laser spark in the gap.

The creation of a laser spark near the tip of an earthed electrode allows a considerable decrease in the time of formation of a stable leader, which starts propagating towards the cloud and captures a part of the laser spark. As a result, its velocity increases compared to the case when the discharge develops in the absence of a laser spark in the gap. This circumstance is of fundamental importance for laser systems of lightning protection. Indeed, owing to its interaction with the laser spark, the leader discharge developing from the tip of the lightning rod may well precede the ascending discharges from the object being protected during its advance towards an approaching descending leader, capture the latter and thus protect the object.

REFERENCES

[1] V. V. Apollonov, A.M. Prokhorov, "Universal laser for industrial, scientific and ecological use", Proceedings of GCL/HPL Conference, St. Petersburg, p. 140, 1998.

[2] V.V. Apollonov, "Ecologically safe High power Lasers", Proceedings of Lasers-2001 Conference, p.3, 2001.

[3] V.V. Apollonov, "High power autonomous CO2 GDL (100kW,CW/P-P modes) for new technologies development and environment protection"; Russia –NATO Int. Seminar, Moscow, 2005, Proceedings of GCL/HPL Conference, Lisbon,p.35,2008.

[4] V.V. Apollonov "New application for high power high repetition rate pulse-periodic lasers", "Laser pulse Phenomena and Applications" Intech,p.19,2010.

[5] V.V. Apollonov, "Oil films elimination by laser", Oboronzakaz, №17, December, p. 33, 2007.

[6] V.V. Apollonov, Yu. S. Vagin, V.V. Kijko, V.I. Kislov "High rep. rate P-P lasers", Patent RF №2175159.

[7] International Electrotechnical Commission Standard: IEC 1024-1 (1990). *Protection of Structures Against Lightning*. Part 1: General Principals.

[8] International Electrotechnical Commission Standard: IEC 1312-1 (1994). *Protection Against Lightning Electromagnetic Impulse (LEMP)*. Part 1: General Principals.

[9] Aleksandrov G.N., Ivanov V.L., Kadzov G.D., et al. *Elektrichestvo* (12), 47 (1980).

[10] Bazelyan E.M., Raizer Yu.P. *Usp. Fiz. Nauk,* 170, 753 (2000).

[11] Aleksandrov G.N., Kadzov G.D., Petrun'kin V. Yu., et al. *Zh. Tekh. Fiz.,* 47, 2122 (1977).

[12] Vasilyak L.M, Vetchinin S.P., Polyakov D.N. *Elektrichestvo,* 1, 59 (1991).

[13] Uchida S., Shimura E., Yasuda H., et al. *Opt. Zh.,* 66 (3), 36 (1999).

[14] Vasilyak L.M., Vetchinin S.P., Doinikov V.A., Mogamedov A.A., Polyakov D.N., Unkovskii S.Yu. *Fiz. Plazmy,* 16, 839 (1990).

[15] Vasilyak L.M., Vetchinin S.P., Polyakov D.N. Proc. XX Intern. Conf. on Phenomena in Ionized Gases (Piza, Italy, 1991) Vol. 3, p.773.

[16] Marin M.Yu., Pil'skii V.I., Polonskii L.Ya., Pyatnitskii L.N., Reingold A.V. *Zh. Tekh. Fiz.,* 57, 1507 (1987).

[17] Olsen J.N., Leaper R.J. *J. Appl. Phys.,* 53, 3397 (1982).

[18] Ozaki T., Miyamoto S., Imasaki K., Nakai S., Yamanaka C. *J. Appl. Phys.*, 58, 2145 (1985).

[19] Dwyer T.J., Greig J.R., Murphy D.P., et al. IEEE Trans. *On Antennas and Propagation,* AP-32, 141 (1984).
[20] Yamanaka T., Uchida S., Shimada Y., et al. Proc. SPIE Int. Soc. Opt. Eng., 3343, 281 (1998).
[21] Kinoshita F., Morooka Y., Uchiumi M., et al. Proc. XII Intern. Conf. on Gas Discharge and their Applications (Greifswald, Germany, 1997) p. 475.
[22] Vasilyak L.M. Materialy konferentsii po fizike nizkotemperaturnoi plazmy FNTP-98 (Proceedings of the Conference on Low-Temperature Plasma Physics) (Petrozavodsk, 1998) Part 2, p. 135.
[23] Danilov O.B., Tul'skii S.A. *Zh. Tekh. Fiz.,* 48, 2040 (1978).
[24] V.V. Apollonov, L.M. Vasilyak et al QE32(2),115-120, (2002)
[25] Parfenov V.N., Pakhomov L.M., Petrun'kin V.Yu, Podlevskii V.A. Pis'ma. *Zh. Tekh. Fiz.,* 2, 731 (1976).
[26] Bunkin F.V., Korobkin V.V., Kurinyi Yu.A., Polonskii L.Ya., Pyatnitskii L.N. Kvantovaya Elektron., 10, 443 (1983) [*Sov. J. Quantum Electron,* 13, 254 (1983)].
[27] Bychkov S.S., Marin M.Yu., Pyatnitskii L.N. Trudy IOFAN, 50, 166 (1995).
[28] Bazelyan E.M., Raizer Yu.P. Fizika molnii i molniezashchity (*The Physics of Lightning and Lightning Protection*) (Moscow: Fizmatlit, 2001).
[29] Apollonov V.V.,Vasilyak L.M. et al Proc. V Intern. Conf. on Phenomena in Ionized Gases (Nagoya, Japan, 2001) Vol.1, p. 259.
[30] Apollonov V.V.,Prokhorov A.M. et al*J. Opt. Soc. Am.* B, 8, 220 (1991).
[31] Apollonov V.V., Prokhorov A.M., Sorochenko V.R., Shakir Yu.A. Kvantovaya Elektron., 15, 1766 (1988) [*Sov. J. Quantum Electron,* 18, 1102 (1988)].
[32] Askar'yan G.A., Rabinovich M.S., Savchenko M.M., Smirnova A.D. *Pis'am Zh. Teor. Eksp. Fiz.,* 1 (6), 18 (1965).
[33] Apollonov V.V. Prokhorov A.M. et al Intern. Forum on Adv. High Power Lasers and Appl. (Osaka, Japan, 1999) p. 63.
[34] Vereschagin I.P., Antsupov K.V., Bazutkin V.V., Beloglovsky A.A., Makalsky L.M., Nikitin O.A., Sysoev V.S., Temnikov A.G. Proc. XXI Intern. Conf. on Lightning Protection (Berlin, Ger- many, 1992) p. 25.
[35] Apollonov V.V., Prokhorov A.M. et al *J. Opt. Soc. Am.* B, 8, å 2, 220 (1991).
[36] Temnikov A.G., Orlov A.V. Elektrichestvo, 8, 18 (1996).
[37] Vereschagin I.P., Temnikov A.G., Koshelev M.A., Makalsky L.M. Proc. IX Intern. Symp. on High Voltage Engineering (Graz, Austria, 1995) p. 2165.

Chapter 18

Multiply Charged Ions and Their Effective Applications

The creation of high-power/energy lasers has opened a new era in the development of basic research and cutting-edge technologies in various fields of practical application. This first, involves the unique properties of laser sources of high-power coherent radiation. They include: a) high monochromaticity (i.e., a small width of the emission line), which offers new opportunities in high-resolution spectroscopy; b) high spatial and temporal coherence (i.e., the occurrence of light oscillations in a coordinated manner, resulting thereby in a distinct interference pattern), which gives a strong impetus to the development of holography and optical information processing methods; c) a relatively high specific energy that can be emitted by the laser, d) a possibility of varying the length of time during which the energy stored in the laser can be emitted in a wide range of durations: from continuous to femtoseconds; e) a possibility of variations in the temporal structure of radiation from tens of Hz up to tens of GHz; and f) a small divergence, which enables tight focusing. Due to these properties, the laser power density, which can affect the substance, reaches a giant value on the order of 10^{20} W/cm^{-2}. Consequently, it is possible to expose a substance to radiation whose power density exceeds all values known today, characterizing natural and artificial sources. This fantastic opportunity has been thoroughly investigated in the recent decades by scientists from different countries. Clearly, by gradually increasing the laser energy and reducing the length of time during which that energy is emitted, it is possible to observe several stages of an interaction like this.

Thus, we first observe a gradual increase in temperature, which will result in heating of almost any substance up to the melting temperature. Then, a further increase in the rate of energy input into the material leads to such a rapid heating that an intensive evaporation starts, bypassing the liquid phase. Under these interaction conditions, one can observe a phenomenon, which seems surprising at first glance: with increasing power density the amount of the evaporated mass of the matter remains virtually unchanged. The explanation of this phenomenon is simple enough. It consists in the fact that as a result of intense evaporation, the material's vapor begins to screen the irradiated surface. Therefore, a significant portion of the laser energy no longer reaches the surface and is absorbed in the vapor produced by laser radiation. In this case, with increasing flux density of the laser energy the vapor temperature should increase, which will lead to the formation of plasma, i.e., ionization of the evaporated material and creation of a plasma screen, preventing a further input of energy into matter. This fact is not surprising and, moreover, extremely high vapor temperatures, up to strong ionization of evaporated material atoms, can be achieved. The temperatures reached due to laser irradiation were so high that scientists had a desire to try to realize the fusion of light nuclei, i.e., to implement a fusion reaction on the scale of a pinhead.

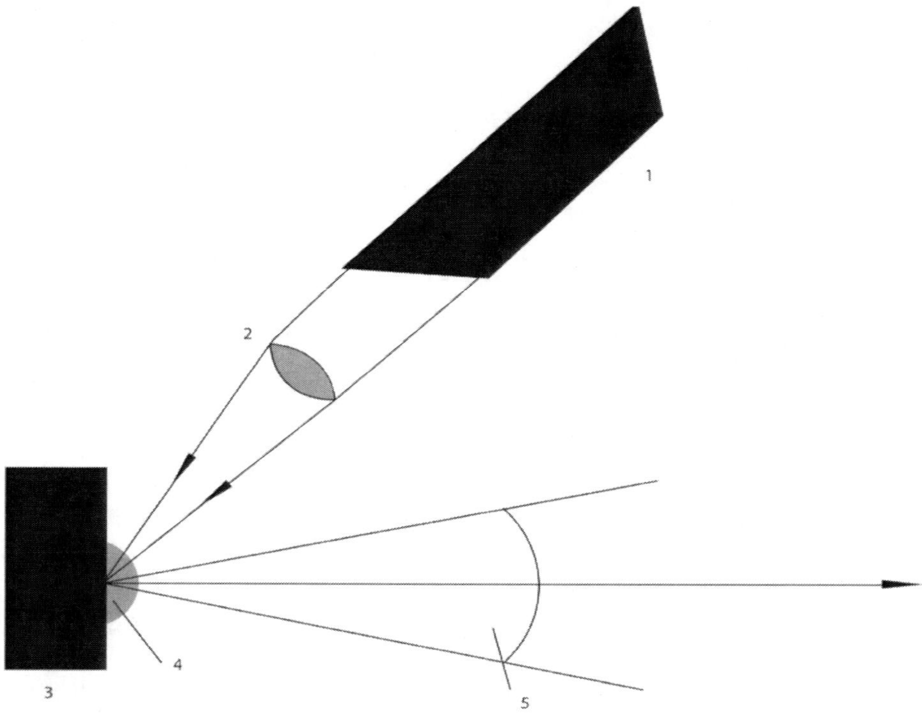

Figure 1. Scheme of a laser-plasma ion source: *1* – laser, *2* – focusing lens, *3* – target, *4* – laser plasma, *5* – ion beam having a maximum charge and maximum energy.

The idea of laser fusion, associated with the ability to create conditions for thermonuclear reactions in the area of laser radiation focusing on a solid target, gave birth to a new area of research, i.e., physics of the interaction of intense laser pulses with matter, whereas the plasma obtained by this method was called "laser plasma". This new direction has progressed dramatically for the last 40 years due to the rapid development of laser physics and engineering. Long-term efforts on the experimental and theoretical study of the properties and characteristics of laser plasma have extended the range of effective applications of laser plasma, originally associated only with laser thermonuclear fusion.

Instead of laser radiation, one can also use high-energy particle beams, which provides a significantly more efficient transmission of the deuterium-tritium fuel energy. A beam of heavy multiply charged ions seems the most promising in this case, because particles with a high charge can be easily accelerated to high energies. Besides, it is easier to obtain higher current values, since the greater the mass, the smaller the influence of the repulsion of ions in the beam due to a space charge. In addition, the beams of heavy multiply charged ions will transfer energy to a target with maximum efficiency. In this case, there arises a question of how to obtain primary beams of heavy multiply charged ions for subsequent acceleration to high energies in accelerator systems. From this perspective, a simple way to implement an inertial controlled thermonuclear reaction is first to obtain the beams of heavy multiply charged ions with the help of high-power laser radiation and then to use these beams for irradiating a deuterium-tritium target in order to reach the fusion temperature. Thus, the methods of obtaining high temperatures and a highly ionized state are of great practical interest.

The main parameters of an ion source are the charge-state distribution of the ions produced and the intensity of the extracted ion beams. Most of the currently existing types of multiply charged ion sources are based on subsequent electron-impact ionization. By the 1970s, heavy-ion accelerators had mainly relied on the use of an ion source based on a reflective (Penning) discharge with oscillating electrons. Ion sources like these exhibited a theoretical limit of charged particles and the intensity of extracted beams, which is determined by the temperature, density and lifetime of the plasma. The electron-beam method of multiply charge ion production was proposed by E.D. Donets in 1967. This ion source, later called an electron-beam ion source (EBIS), was conceived as a pulsed source to produce multiply charged ions for high-energy ion accelerators. To increase the charge of the ion produced by an EBIS, the energy of ionizing electrons should be equal to 100 – 200 keV, and the retention time of the ions should amount to tens of seconds or even minutes. The EBIS uses a linear electron accelerator operating in the continuous mode. Ion sources with a short electron beam having a length of less than 10 cm formed a separate line of research and were called an electron-beam ion trap (EBIT). During the 1970s and 1980s the EBIS's were used to produce record-high charge states of heavy ions. Using a KRION-2 cryogenic electron-beam ionizer, scientists at the Joint Institute for Nuclear Research (JINR) (Dubna, Russia) managed to produce highly charged xenon ions, while using an EBIT, researches at the Lawrence Livermore National Laboratory (LLNL) (Livermore, USA) produced highly charged uranium ions. The main disadvantages of the EBIS in comparison with other types of sources are ion losses and relatively low intensity (10^{10} s^{-1}) of the produced ions. The most widely used sources of multiply charged ions for accelerators and nuclear physics are electron-cyclotron resonance (ECR) sources. An ECR source is an open magnetic trap for plasma confinement. Electrons and ions are generated as a result of electron impact ionization. In turn, the electrons resulting from the ionization of neutrals and ions are heated to energy of several keV by the microwave radio-frequency field, whose frequency is equal to the Larmor frequency of the electron spin in the longitudinal magnetic field of the trap. Increasing the degree of ionization in the ion source is the result of successive ionization during the holding period of the ions. To date, all large accelerator centers in the world are equipped with such ion sources. However, modern requirements to the sources significantly exceed their capabilities. The first step is to increase the ion beam current because ESR sources are promising for continuous ion accelerators due to their ability to generate rather intense (up to 10^{13} s^{-1}) beams of medium and heavy ions in continuous mode.

In contrast, a laser-plasma generator of multiply charged ions produces a large number of heavy ions in the mode of short periodic pulses, which is of interest for ion accelerators operating in pulsed mode. The source of this type is also promising for research in the field of heavy-ion fusion. The principle idea of a laser-plasma source of multiply charged ions was proposed 45 years ago together with author of the book by researchers at the Moscow Engineering Physics Institute (MEPhI). The laser-plasma generator is based on the physical phenomenon of generation of highly ionized states of atoms under irradiation of the surface a solid target by a high-power, focused laser pulse [1]. When expanding into a vacuum, high-temperature laser plasma produces a powerful stream of charged particles. The advantages of a laser-plasma source as compared to other types of ion-pulse generators include:

- ability to generate multiply charged ions of almost any elements of the periodic table, and
- the ability to generate an intense, short (1 – 100 μs) ion pulse with a record-high brightness.

The main difference of the laser plasma from other high-temperature plasma objects is a high energy density in matter, caused by the ability of lasers to concentrate in a short time the energy in small volumes of material (10^{-6} cm^{-3}). Changing the power and wavelength of heating radiation allows one to control the temperature and the laser plasma density. Irradiation of a target surface by focused laser radiation with a power density >10^8 W/cm^2 produces a plasma plume with a large temperature and density. Depending on the laser radiation parameters, the electron temperature in the plume can be as high as T_e = 2.10 keV at an electron density of 10^{19} – 10^{21} cm-^{3}. This allows one to produce highly charged states of ions in the laser plasma. Such characteristics of the ion component of the laser plasma as highly charged states, intensity and type of angular distribution are exceptionally favorable for its use as a source of multiply charged ions for accelerators. The first laser-plasma generator for an accelerator complex was constructed thirty years ago (joint work of MEPhI and JINR researchers). Requirements imposed by accelerators to ion sources (pulse repetition rate, up to 10 Hz; stability of the output parameters of the ion beam, ± 10%; and time of continuous operation) set strict limits as to the choice of the laser type and its maximum achievable energy. Due to the development of laser technology in the last 30 years, transverse discharge pumped CO_2 lasers seem most attractive for the use in laser-plasma generators. The output energy of these lasers can vary from 1 to 1000 J at laser pulse durations from 0.01 μs to 1 μs and a pulse repetition rate up to 10 Hz. Using CO_2 lasers to generate multiply charged ions from laser plasma, due to their high level of technical development, relative ease and low cost even in the configuration of a low-frequency repetitively pulsed facility with a long running time, is currently very promising. An increase in the repetition rate of high-power and short laser pulses up to 30 – 50 kHz will significantly increase the yield of multiply charged ions and dramatically enhance the efficiency of laser-plasma generators.

Thus, evaporation of the material and heating of the vapor lead to a plasma state, i.e., when vapor is heated, electrons are detached from atoms due to a temperature rise. In this case, the higher the temperature, the more electrons are 'torn away' from an atom. Therefore, an ionized vapor, i.e., plasma, consists of electrons and ions. If we assume that the temperature of the electrons and ions is the same, and since the electrons are much lighter than the ions, they travel much faster. Due to such a high velocity, the electrons in the process of the gas-dynamic expansion will be the first to fly out of the plasma region. Then, at the forefront of expanding plasma there occurs separation of the negative charge of the emitted electrons and the positive charge of the ions. In this case, due to the Coulomb interaction, the electrons pull the ions. This process will lead to the separation of electrons and ions in space and time (quenching effect), which interferes with the recombination of electrons and ions in a laser plasma. Therefore, a directional flow of highly charged ions is produced in the form of a beam that propagates in a direction normal to the target surface. For ions with a maximum charge and energy, the angle of their departure decreases, and one can observe the effect of self-focusing of ions, depending on the charge multiplicity. In this case, the interaction of a focused laser beam with a high energy flux density gives us a very simple and effective source of multiply charged ions without the need for any pulling or focusing fields.

Figure 1 shows a scheme of a laser-plasma source. In this laser-plasma source of ions, the ionization degree is controlled by the power density of laser radiation and can reach a few dozen; this means that it is possible to completely ionize a significant part of the elements of the periodic table. It should be noted that prior to the advent of lasers it was practically impossible to achieve the degree of ionization $Z > 10$ and thus, for example, to measure the bonding energy of the electrons in the atom. We emphasize that in this case new practical and promising applications, rather than purely fundamental results, proved to be important.

A laser-plasma source of ions and nuclei is promising for charged particle accelerators. In January 1976, MEPhI and JINR scientists realized for the first time in the world the acceleration of carbon nuclei from the laser-plasma source (Figure 2). Carbon nuclei from the laser-plasma source were introduced into a linear accelerator and accelerated in the synchrotron channel to energy of 50 GeV. If previously the accelerator accelerated singly charged protons, the use of multiply charged ions made immediately increased the energy of accelerated particles by Z times. This made it possible to make the next step in obtaining relativistic beams of complex nuclei. In recent years, such work has been carried out at CERN (Switzerland) to obtain multiply charged ions of heavy atoms with $Z > 50$ by using a laser-plasma source of ions.

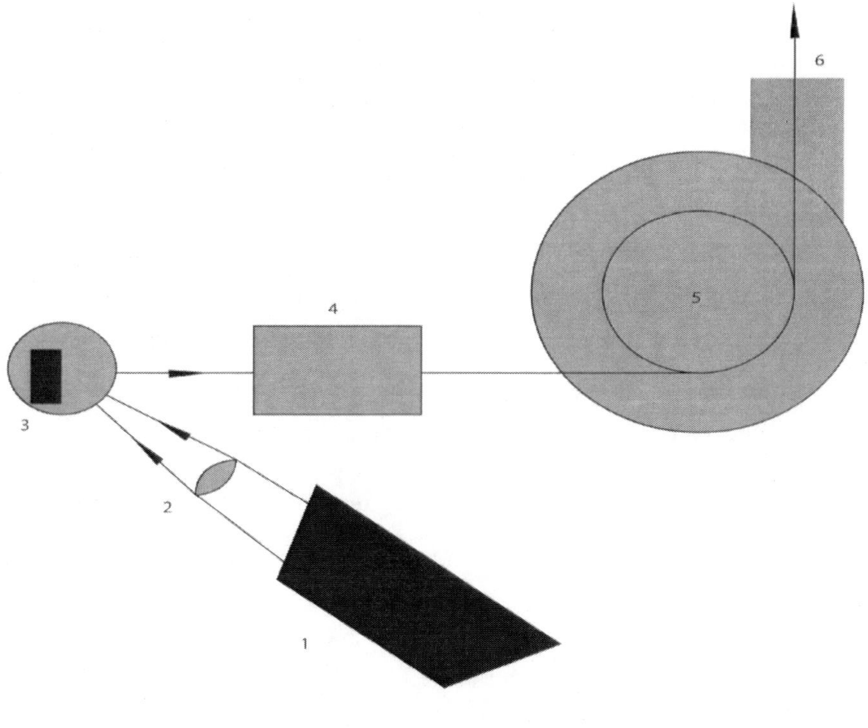

Figure 2. Scheme of the acceleration of carbon nuclei at the proton synchrotron using a laser-plasma source of carbon nuclei C^{6+}: *1* – laser, *2* – lens, *3* – graphite target, *4* – linear accelerator, *5* – synchrotron, *6* – output beam.

Nowadays, laser plasma finds numerous applications in various fields of fundamental physics: X-ray spectral analysis of multiply charged ions, high energy density physics and physics of shock waves, modeling of space-physical and high-temperature processes, X-ray lithography, etc. In recent decades, ion sources have been intensively designed and developed. An incentive to create them stems from the need of heavy-ion accelerators. The ion source is the first element of an accelerating complex, which determines the structural characteristics, parameters, efficiency and capabilities of the accelerator. Generation of high-current beams of multiply charged ions is one of the most promising areas of research, which finds wide application in science and technology. As an example of employment of such beams, we can single out the following:

- a source of multiply charged ions of heavy elements in the fore-injectors of particle accelerators;
- a source of cluster ions and molecules;
- a source of beams of highly ionized atoms to measure the interaction cross sections in such field of atomic physics as nuclear fusion, ionosphere physics, astrophysics and investigation of the laser plasma;
- production of uniform films to form the fine structure of multilayer X-ray mirrors and diamond films;
- radiation material science, ion implantation to alloy semiconductors or change the surface properties of materials; and
- formation of ion beams with a specific charge state in order to create a gain medium for the X-ray lasers.

Recent years have seen a significantly growing interest in a wide practical use of sources of multiply charged heavy ions. Thus, in Dubna (JINR) scientists have developed a unique technology for producing filters. In a polymer film a passing heavy ion leaves a channel having large radiation damage, and complex molecules are broken down into smaller parts, i.e., radicals. For this reason, the region of the track (channel) becomes sensitive to chemical reactions. After etching the polymer film exposed to heavy ions, through channels are formed. The diameter of these channels depends on the temperature and etching time, and can range from five nanometers to tens of microns. Such channels can be used in biology and medicine for separation of different types of viruses and bacteria (having dimensions greater than 0.2 μm), and the protein molecules (enzymes); in addition, they can be used in semiconductor technology and for water filtration.

Heavy-ion beams are quite promising for radiotherapy, in particular for the treatment of cancers. The use of laser-plasma beams of heavy ions is much more effective in comparison with the use of large and expensive installations, because they make the radiotherapy of limited areas of the human body possible due to the large ion charge. Heavy ion sources are also needed for the synthesis of transuranic chemical elements, lying on the so-called 'islets' of stability, which are predicted in the region $Z = 114$ and $Z = 126$.

No less interesting and practically important application of laser-plasma ion sources is ion implantation of metals. It has been shown experimentally that the bombardment of the surface of metals by ions can significantly change their physical properties. The reason for this is, firstly, the effect of the introduction of these or other ions, that is, the change of the elemental composition by doping, and secondly, the change in the structure of the surface layers. Such methods have been developed since the 1970s; they have shown that ion implantation can significantly alter the physical properties of metals, including corrosion resistance, surface friction, wear resistance, etc. It should be emphasized that the laser-plasma implanter for modification of metal surfaces is attractive due to the fact that it has a small size, which makes it very easy to use this or that type of ions and to obtain multi-element ion beams of required stoichiometric composition. Besides, one can easily obtain a pulse current of the order of an ampere and, using a standard frequency laser, have reasonable time values to ensure the required irradiation dose.

The interaction of laser radiation with matter is another very important and promising sphere of analytical research, whose aim is to determine the elemental composition of the material irradiated by the laser. This physical prerequisite of this phenomenon were the following experimental facts: the possibility of evaporation of any material regardless of its composition, nature and physicochemical properties, i.e., laser radiation at appropriate energy and power parameters can evaporate metals, semiconductors, textiles, dielectrics, ceramics, biological object (living tissue, blood), etc. And, as mentioned previously, due to the plasma shielding an almost constant quantity of substance (about one microgram) is evaporated. The next important factor is that the evaporated material is almost completely ionized in the same laser beam, and the ions are formed into a beam directed along the normal to the laser-irradiated targets. Thus, a laser-plasma ion source also allows registration and separation of particles in electric and magnetic fields.

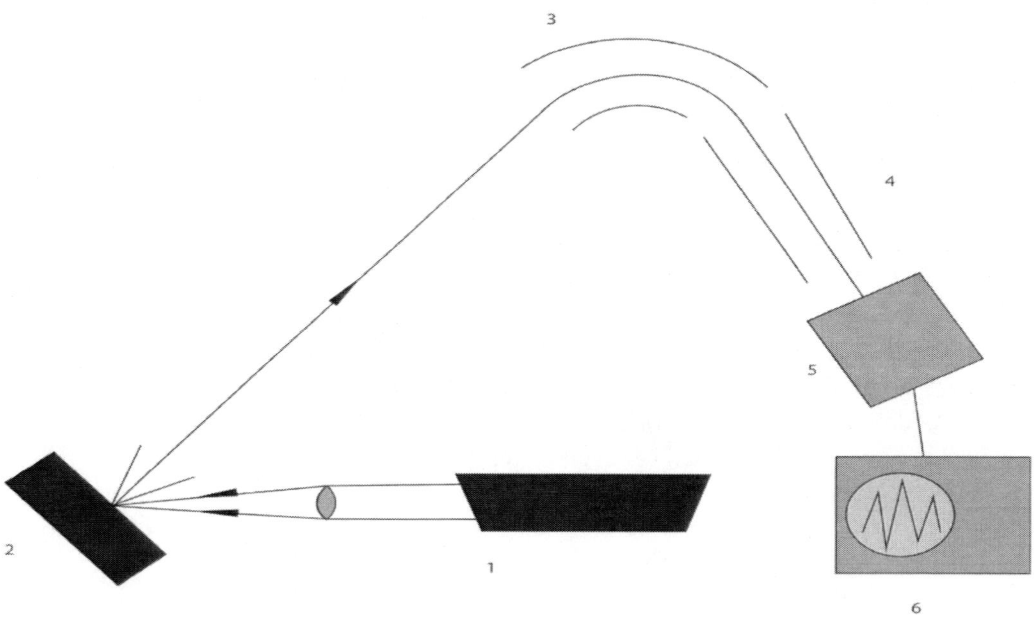

Figure 3. Scheme of a time-of-flight mass spectrometer with a laser-plasma ion source: *1* – laser, *2* – irradiated target, *3* – electrostatic analyzer, *4* – drift length, *5* – registration system of ions, *6* – oscilloscope.

Figure 3 shows the scheme of a time-of-flight mass analyzer using a laser-plasma ion source. The simplest scheme involves a combination of laser-plasma ion sources with an electrostatic analyzer and time-of-flight mass separation. Figure 3 shows the principle of operation of the laser device for express elemental analysis. Radiation from the laser is focused on a target, the elemental composition of which we wish to identify. Upon arrival of the laser pulse, a part of the target is vaporized and ionized. The main portion of the ions is formed into an ion beam which is directed along the normal to the surface of the target. In this case, the ion beam corresponds to the target composition. The electrostatic analyzer identifies the ion beam, ideally with a single energy value. Then, this monochromatic beam of ions, formed from elements with different masses, will be separated in time, i.e., the lightest ions will first reach the registration system. From the registration system the electrical signal is fed to an oscilloscope where a certain maximum weight of a chemical element corresponds to each maximum. This time-of-flight mass spectrometer with a laser-plasma ion source allows one to analyze in real time a wide class of materials with a sensitivity of $10^{-3} - 10^{-4}$ atomic percent with the mass resolution of ~300.

Very useful and promising was the use of a laser-plasma ion source in the well-known scheme of a spectrometer with double focusing, which combines an electrostatic analyzer with a mass spectrometer sweeping the magnetic field. In this case, the ion beam from the target being irradiated passes through the electrostatic analyzer, where the beam is monochromatized and then falls into a uniform magnetic field, where, as is known, ions are separated according to their charge-to-mass ratio. Because use is made of singly charged ions, the beam of ions with different masses is swept by the magnetic field and recorded, for example, on a photographic plate. Such a device exhibits high analytical characteristics. The mass resolution reaches the value of up to 5000, which means that it is possible to detect the isotopes of all elements of the periodic table and register all the elements from hydrogen to uranium. This device also makes it possible to detect impurities in the material, when the concentration is about $10^{-5} - 10^{-6}$ at.%.

A distinguishing feature of the operation of a mass spectrometer with a laser-plasma ion source, except for its versatility and analysis of any substance, consists in the fact that all the elements are registered simultaneously, and knowing the evaporated mass one can perform an analysis without any standards, if use is made of natural isotope ratios of different elements. These characteristics determine the possible applications of mass spectrometers with laser-plasma ion sources, which include not only the control of impurities in metals and semiconductors, but also opportunities for environmental pollution control in the air and in the soil. Thus, when measuring the composition of wheat grains collected closer than 300 meters from heavily loaded transport highways, lead is found. Besides, in pearl lipsticks mercury is detected. A large set of heavy metals is found in analysis of the scales of fish caught in the water near untreated sewage into the river,.

Using such devices in biology and medicine also undoubtedly holds promise for medicine and forensics. Thus, it is possible to carry out an analysis of the elemental composition of blood using a piece of gauze soaked with blood or a substrate with a drop of blood on it [2]. For example, to forensics, it is important that the content of heavy elements in a person's blood is strictly individual [3]. In analyzing biological tissue elements one can detect drastic changes in the number of different elements, such as Ca, Cl, Si, Al, etc., which indicates certain diseases.

One cannot but mention one more very important and interesting application of laser plasma. Because a high degree of ionization can be achieved in laser plasma, at the early stages of the expansion the laser plasma becomes an intense source of soft X-rays that are of bremsstrahlung and recombination nature; therefore, we deal with the radiation arising due to changes in the velocity during the interaction of charged particles and the recombination of electrons and ions [4]. In this case, the wavelength of the recombination radiation is determined by the degree of ionization and can be controlled both by the laser parameters and by using materials with different ionization potentials. Soft X-ray radiation can be focused and thereby beams with high energy characteristics can be generated.

Such a laser-plasma source is very promising for use in X-ray lithography in the manufacture of semiconductor chips. Due to the small wavelength and small focusing spot size one can significantly improve the resolution of the photomask used in the production of semiconductor chips. In addition, a laser-plasma soft X-ray source enables the creation of a microscope working in the soft X-ray range, i.e., using photons with such a low energy that they will not cause genetic mutations and destroy the structure of a living cell, but will allow one to follow the behavior of individual chemical elements.

Another very promising technological application of laser plasma is the production of thin films of complex composition. In this case, use is made of the flows of the plasma produced in the interaction of matter with radiation of a Q-switched laser, i.e., in the mode of high-rate energy input into the matter. Thus the target can be made of different materials determining the desired stoichiometry of the resultant film. In this case, high rates of energy input, as noted above, allow one to preserve the stoichiometric composition of the target and to produce complex compounds on the substrate. The use of frequency lasers makes it possible to deposit materials epitaxially, i.e., in small portions, layer by layer, while maintaining the crystalline structure. In addition to producing complex semiconductor compounds, this technology has been very useful in creating high-temperature multicomponent superconductor compounds.

This technology is characterized by the fact that complex compounds in the form of single crystals grow on substrates. We can assume that one of the reasons for this is the concomitant recombination radiation from the laser plasma, which activates electron shells of atoms during crystallization, which stimulates the growth of crystals with more energy-intensive orientations. Experimentally, many laboratories have shown that by vaporizing carbon by laser radiation from a graphite target, it is possible to produce thin diamond-like films and more complex structures in the condensation process under certain conditions.

Note, in conclusion, that the sphere of possible applications of multiply charged ions produced in the interaction of high-power pulsed and repetitively pulsed laser radiation with matter is expanding every day. The next step is the introduction of high-power repetitively pulsed laser systems with high pulse repetition rates, providing a solution to the problem of a plasma screen and making it possible to deliver much higher average power flows into the matter, thereby producing more powerful beams of multiply charged ions for their effective use.

REFERENCES

[1] V.V. Apollonov, Yu.A. Bykovskii, N.N. Degtyarenko, V.F. Elesin, Yu.P. Kozyrev, S.M. Sil'nov. Production of multiply charged ions by interaction between a powerful laser pulse and a solid. *Pis'ma Zh. Eksp. Teor. Fiz.*, 11, 377 (1970) [*JETP Lett.*, 11, 252 (1970)].

[2] V.V. Apollonov, Yu.A. Bykovskii, Yu.P. Kozyrev, S.M. Sil'nov, T.N. Filinskii. Possibility of using mass spectrometry for elemental analysis of blood, Report No. 15/17-75, Moscow Engineering Physics Institute, Moscow, 1970.

[3] V.V. Apollonov, Yu.A. Bykovskii, Yu.P. Kozyrev, S.M. Sil'nov, T.N. Filinskii. Possibility of using mass spectrometry in forensic science, Report No. 15/17-73, No. 15/17-74, No. 15/17-76, Moscow Engineering Physics Institute, Moscow, 1970.

[4] V.V. Apollonov, K.Kh. Kazakov, N.V. Pletnyev, V.R. Sorochenko. Current measurements in the plasma produced by train of 10.6 μm sub-nanosecond laser pulses. *Proc. SPIE*, 3886, 198 (2000).

Conclusion

High technology development is a major topic of many politicians, who fully understand (and this point of view is supported by many experts) that laser technology is one of the most effective ways to gain technological superiority. As a matter of fact, laser technology today makes it possible: to provide a new level of development of industry, science and technology of your country; to revive scientific and technological strength of the country; to make your country a leader in the field of technical, scientific and technological progress; to revive a large number of enterprises in many sectors of industry, which have been well known for their developments in previous years; to strengthen your country's leading position in space and to ensure strategic and geopolitical priorities in the modern world; and to derive benefit from the sale of a wide range of laser programs, which is today comparable to the sale of traditional products and resources. The future of laser technology is very bright!

About the Author

The author of the book is V.V. Apollonov, Academician of the Russian Academy of Natural Sciences and Academy of Engineering Sciences, Doctor of Physics and Mathematics, Professor, General Director of Energomashtehnika Ltd., Head of the Department of High-Power Lasers at the A.M. Prokhorov General Physics Institute, Russian Academy of Sciences. In 1982 and 2002, he was awarded the State Prize of the USSR and Russia, respectively. V.V. Apollonov is a leading expert in the field of high-power/energy laser systems and interaction of high-power/energy laser radiation with matter. Being the author of over 1770 publications (14 books, chapters 27, 546 presentations and 147 patents, 954 articles, inf. from Research Gate), V.V. Apollonov is widely recognized by the scientific community. He is the organizer of international conferences. More than thirty of his disciples have become candidates and doctors of physical and mathematical sciences under his supervision. V.V. Apollonov is a member of many international organizing committees of major conferences and symposia in the field of high-power/energy lasers and their applications. In addition, he is a member of the Presidium of the Russian Academy of Natural Sciences. He has repeatedly acted as a keynote speaker in international scientific forums.

Email: vapollo@rambler.ru.

INDEX

#

A

B

C

F

G

H

I

J

K

L

M

N

O

P

Q

R

S

T

U

V

W

X

Y